T5-AFF-888

CHANGING REGULATION

CONTROLLING RISKS IN SOCIETY

Related Elsevier Books

FRICK	Systematic Occupational Health and Safety Management
FULLER & SANTOS	Human Factors for Highway Engineers
HALE & BARAM	Safety Management and the Challenge of Organisational Change
HALE, WILPERT & FREITAG	After the Event: From Accident to Organisational Learning
HARTLEY	Managing Fatigue in Transportation
HAUER	Observational Before-After studies in Road Safety
SOARES	Advances in Safety and Reliability

Related Pergamon journals – sample copies available online from:
http://www.elsevier.com

Accident Analysis and Prevention
Editor: Frank Haight

Journal of Safety Research
Editor: Thomas W. Planek

Safety Science
Editor: Andrew Hale

Reliability Engineering and System Safety
Editor: G.E. Apostolakis,

Structural Safety
Editor: B. Ellingwood

CHANGING REGULATION

CONTROLLING RISKS IN SOCIETY

Edited by

BARRY KIRWAN
Eurocontrol, Brussels and Paris

ANDREW HALE
Delft University of Technology, The Netherlands

ANDREW HOPKINS
Australian National University, Canberra, Australia

2002

Pergamon

An Imprint of Elsevier Science

Amsterdam – Boston – London – New York – Oxford – Paris
San Diego – San Francisco – Singapore – Sydney - Tokyo

ELSEVIER SCIENCE Ltd
The Boulevard, Langford Lane
Kidlington, Oxford OX5 1GB, UK

First edition 2002

Library of Congress Cataloging in Publication Data
A catalog record from the Library of Congress has been applied for.

British Library Cataloguing in Publication Data
A catalogue record from the British Library has been applied for.

ISBN: 0 08 044126 2

♾ The paper used in this publication meets the requirements of ANSI/NISO Z39.48-1992 (Permanence of Paper).
Printed in The Netherlands.

CONTENTS

LIST OF CONTRIBUTORS

Michael Baram is Professor of Law and Director of the Center for Law and Technology at Boston University School of Law where he directs research and teaches Environmental Law, Products Liability Law, and Biotechnology Law & Ethics. He also holds joint appointments with the School of Public Health and the Bioinformatics Department at Boston University. He graduated from Tufts University School of Engineering and later received his law degree from Columbia University School of Law. He has served on several National Academy of Sciences Committees and various expert commissions and advisory boards, and has consulted for many governmental and private organisations regarding various regulations, liabilities and other aspects of agency and corporate risk management. For over 25 years, he was a partner in the Boston law firm of Bracken & Baram. He has published numerous articles and several books including *Alternatives to Regulation* and *Safety Management* (with A.Hale).
Email: mbaram@bu.edu

Gerhard Becker has a degree in mechanical engineering and gained his experience in safety analysis by studying nuclear accident scenarios for various nuclear power plants in Germany and by participating in the international reactor safety experiments in Marviken, Sweden. The influence of human behaviour on the safety of complex technical systems has been the main focus of his work over the last 20 years. His main interest recently has been directed to the relevance of organisation and management for the safety of industrial technology. His work is based on comprehensive experience gained as head of an interdisciplinary team concerned with the ergonomic design of technical systems, which is seen as the precondition for the safe performance of human operators. He is currently project manager at the Institute for Nuclear Technology and Radiation Protection (IKS) of TUV Rheinland/Berlin-Brandenburg. This company, a Technical Inspection Association with over 125 years of tradition and experience, is working in all areas of technical safety, documented quality and environmental protection.
e-mail: Gerhard.Becker@de.tuv.com

Geerda van Gaalen has worked at the Academic Medical Center of the University of Amsterdam since 1991, when she was invited to join the staff of the Division of Cardio-thoracic Surgery as administrative assistant with a special focus on research and education.
Ms. van Gaalen has coauthored several articles on the subject of safety in health care.
She is an active member of various regional and national committees tasked with organizing specialized training sessions for nursing staff, residents in cardio-thoracic surgery, and certified cardio-thoracic surgeons. Ms. van Gaalen has organized numerous international congresses.

A long career in the medical field has provided her with extensive international work experience in university hospitals located in Rotterdam, Cape Town, and Ottawa.

Louis Goossens holds a degree in chemical engineering and a Ph.D. in technical sciences from Delft University of Technology, the Netherlands. He is associate professor in safety science at Delft University of Technology, Faculty of Technology, Policy and Management. His main field of interest is risk assessment and developing and applying tools for qualifying and quantifying technical risks in the broader context of developing safety management systems. He was coordinator of the Expert judgement project for developing methods for the Dutch Ministry of Housing, Physical Planning and Environment since 1985. A Procedures Guide for Structured Expert Judgement was published in 2000 (EUR 18820). He was also project manager of the EC/USNRC joint project of the Commission of the European Communities and the United States Nuclear Regulatory Commission on Uncertainty Analysis of Probabilistic Accident Consequence Codes (expert judgement studies to establish the uncertainty distributions over the major parameters in the consequence codes and to perform uncertainty analyses using these expert judgement data). He is currently working on risk assessment projects for the chemical industries, aviation industry and railway industry.

Harry Gundlach studied physics and mathematics in Utrecht, the Netherlands and in Oxford, UK. For more than 15 years he worked as Manager Research and Development in the electrical industry (HOLEC), in The Netherlands. He was General Manager of the Dutch Council for Certification (RvC) and the Council for Accreditation (RvA). In May 1998 he resigned as General Manager of RvA to become its Manager International Affairs. He lectures part-time on quality, certification and standardisation at universities in the Netherlands. He is also active as a lead assessor and performs audits based on the accreditation standards. Harry Gundlach was one of the founders of the "European Accreditation of Certification" and the "International Accreditation Forum", of which he was chair from 1996 to 1998. He is a member of the Board of Administrators of the European Organisation for Conformity Assessment, (EOTC). He was a member of the "World Quality Council" (WQC) and is a 'Distinguished Fellow' of the Institute of Directors in India. He is chairman of ISO/TC 67/WG 2, active in providing guidance on principles and application of conformity assessment in the petroleum and natural gas industry.
Email: Harry.Gundlach@rva.nl

Andrew Hale is professor of Safety Science at the Delft University of Technology in the Netherlands. He has worked in the area of safety since 1966, starting with research on human factors and accident causation and moving on later to safety management and safety regulation. His PhD was on the role and training of factory inspectors in the UK. Since his move to the Netherlands in 1984 he has carried out studies on the regulation and assessment of safety management systems and occupational health and safety advisory services. He led a team assisting the Dutch mines inspectorate, responsible for the regulation of the oil and gas

extraction industry, to change from a technically oriented inspection approach to one centring more on management systems - the work described in this book. He has also worked in the areas of airport and maritime safety regulation and management. He is chairman of the scientific advisory board for the Dutch Road Accident Research Institute and has served on a number of review boards for safety research institutes in the Netherlands, Norway and France. He is editor of the journal *Safety Science*.
Email: a.r.hale@tbm.tudelft.nl

Jan Hovden is Professor of Safety Management at the Norwegian University of Science and Technology, Department of Industrial Economics and Technology Management, Trondheim, Norway. He was a member of the governmental commission on "A Vulnerable Society" (1999-2000) and a member of a study group on "Decision-making under Uncertainty" at the Norwegian Academy of Science and Letters, Oslo (2000-2001). He has a degree in Political Science from the University of Oslo.
Email: hovden@iot.ntnu.no

Andrew Hopkins is Reader in Sociology at the Australian National University in Canberra. He is the author of numerous articles on the management and regulation of occupational health and safety, as well as several books. *Making Safety Work: Getting Management Commitment to Occupational Health and Safety* (1995); *Managing Major Hazards: the Lessons of the Moura Mine Disaster*, (1999); *Lessons from Longford: The Esso Gas Plant Explosion*, (2000). He was an expert witness at the Royal Commission into the causes of the fire at Esso's gas plant at Longford in Victoria in 1998. He recently served as an expert member of a Board of Inquiry into the exposure of Air Force maintenance workers to toxic chemicals. He speaks regularly to industry groups and at conferences about the lessons of Longford, and of Moura. He has a BSc (Hons) and an MA from the ANU, and a PhD from the University of Connecticut.
Email: Andrew.Hopkins@anu.edu.au

Barry Kirwan studied Psychology at Southampton University, immediately followed by a Masters in Work Design and Ergonomics at the University of Birmingham (UK). He then moved into consultancy (HRA Ltd) as a research scientist working on nuclear power plant risk assessments, specialising in human reliability assessment. He became the founder member of Technica's Human Factors Unit, working primarily in drilling and offshore risk assessments, including marine risk assessments for offshore platforms. He was briefly a director of HRA Ltd, diversifying the company's projects into the offshore area and other areas. He then moved to British Nuclear Fuels and became responsible for human reliability in the design of the THORP project, one of the largest construction projects at the time. As the design phase drew to an end, he moved to Birmingham University, where he lectured Human Factors and Engineering students, also carrying out research and development in nuclear and offshore human reliability assessment. He then became Head of Human Factors for National Air Traffic

Services, the principal Air Traffic Control organisation in the UK, managing Human Factors staff and contractors. He extended the traditional HF work in NATS to consider human error and various safety aspects of new and existing projects and systems, as well as feeding more human factors into the design of new systems. His most recent move has been to Eurocontrol, in Brussels and Paris, working on a number of projects concerned with the future automation tools aimed for 2005 onwards. He has published two textbooks (on task analysis and human reliability assessment) and various articles.
Email: barry.kirwan@eurocontrol.fr

Tore J. Larsson is Professorial Fellow, Monash University Accident Research Centre, Melbourne, Australia. He has a PhD (Industrial Ergonomics) from the Royal Institute of Technology, Stockholm and an MA (Social Anthropology) from University of Stockholm. He did research at the Institute of Applied Psychology, University of Stockholm in perceived difficulty and strain and at the Laboratories for Clinical Stress Research, Karolinska Institute, Stockholm, in stress and retirement from work. Between 1979-1985 he was one of a multi-disciplinary team building the Occupational Accident Research Unit, Royal Institute of Technology, Stockholm. He founded the Institute for Human Safety & Accident Research (IPSO), Stockholm, together with the Swedish Labour Market Insurances, extending its work later to Australia and New Zealand. He developed the National Injury Information System for severe occupational injuries in Sweden (TSI). His main interest areas are accident and injury analysis, criteria for prevention, occupational risk assessment, and the implementation of worksite measures. He is responsible for the bi-annual national screening of severe occupational injuries with the Labour Market Insurances in Sweden and he has recently built the occupational risk monitoring system with the Victorian workers' compensation insurance in Australia. He is the author of more than 100 published articles and numerous training films.
Email: Tore.Larsson@general.monash.edu.au

David Maidment OBE, BA, MCIT, MIOSH, worked for 34 years for British Rail and 2 years for Railtrack, retiring in 1996. He was Chief Operating Manager for BR's London Midland Region in the mid-1980s, followed by spells at the BR Board HQ as Quality & Reliability Manager and Head of Safety Policy. He became Controller, Safety Policy for Railtrack at that company's inception and became a part time Director of International Risk Management Services, a safety management consultancy specialising in rail transport, after his retirement. He has undertaken safety consultancy work for railway companies and regulators in the UK, Ireland, Hong Kong, Canada, South Africa, Australia and New Zealand during the last 10 years and is currently a member of Railtrack's panel of independent accident inquiry chairmen. He is also Chair of Amnesty International UK's Children's Human Rights Network, Co-Chair of the UK Consortium for Street Children and Founder/Chairman of the Railway Children, an industry charity working for street children internationally.
Email: Maidmentrail@aol.com

Bas de Mol, MD, PhD, JD is chief of the division of Cardio-thoracic Surgery at the Academic Medical Centre of the University of Amsterdam. He has been part-time professor of Safety Science and Systems Engineering in Health Care at the Delft University of Technology, and is now part-time professor of Biomedical Technology at the Eindhoven University of Technology. He is an expert in the field of risk and damage assessment, particularly in the area of health care.
Email: basdemol@hetnet.nl

Leena Norros is working at the Technical Research Centre of Finland in the Department VTT Automation where she leads a research group in Human
Factors. She also acts as an associate professor at the Helsinki University and teaches work psychology. With her group she has studied work in various process control domains with the aim of developing a contextual ecologically oriented approach in the analysis of decision-making and expertise in dynamic, complex and uncertain situations.
Email: leena.norros@vtt.fi

Joy Oh has degrees in medical biology and chemical technology. He is a policy co-ordinator with the Ministry of Social Affairs in the Netherlands. This ministry is responsible for the laws, regulations and policy with respect to safety and health of workers. Part of the ministry is the Labour Inspectorate. Apart from making policy the ministry is also involved with the development of codes, standards and support tools with respect to auditing and inspection of major hazard installations and safety management systems. Joy. Oh is a member of the Commission of Competent Authorities established under the Seveso II directive. This committee acts as an advisory committee for the EU Commission.
Email: JOh@minszw.nl

Ibo van de Poel is assistant professor (UD) at the Department of Philosophy, Faculty of Technology, Policy and Management, Delft University of Technology in the Netherlands. He graduated in the Philosophy of Science, Technology and Society at the University of Twente, where he also obtained his PhD in science and technology studies in 1998 with a thesis entitled "Changing Technologies A Comparative Study of Eight Processes of Transformation of Technological Regimes". He lectures and does research in the field of engineering ethics and the ethical and social aspects of technology. At the moment, his main research activities are related to ethical issues in technological risk and in engineering design. He has published in several international journals (http://www.tbm.tudelft.nl/webstaf/ibop).
Email: i.r.vandepoel@tbm.tudelft.nl

Teemu Reiman holds a degree in psychology from the University of Helsinki. He works as a research scientist with the Technical Research Centre of Finland's (VTT) Human Factors research group. He is interested in industrial psychology and organisational culture in particular. He is also preparing a doctoral thesis on the subject of organisational culture in high reliability organisations.
Email: teemu.reiman@vtt.fi

Dr.-Ing Jürgen Walther has been head of the Department for Supervision of all Bavarian nuclear power plants in the Bavarian State Ministry for State Development and Environmental Affairs for the last 13 years, with a break of two years when he headed the Department "Basic Issues, Coordination, New Technologies" inside the Nuclear Power and Radiation Protection Division. He is a graduate of the Technical Universities of Munich and Hannover. He worked for seven years as Technical Civil-Servant in the Federal Railroad Operation (Deutsche Bundesbahn) before moving to the Bavarian Ministry for State Development and Environmental Affairs in 1978. Before becoming head of the Department of Supervision he was responsible for licensing procedures for nuclear power plants and for radiation protection and ecology, particularly relating to follow-up measurements in the aftermath of the Chernobyl disaster.

Jeremy Williams, BSc, MSc, is a Principal Inspector in the Nuclear Safety Directorate of the Health and Safety Executive. He has 32 years' human factors engineering experience in a range of organisations including the British Railways Board, Rank Xerox Engineering Group, the United Kingdom Atomic Energy Authority, Nuclear Electric, DNV Technica and Electrowatt Engineering Services (UK) Ltd. Mr Williams is the originator of the HEART (Human Error Assessment and Reduction Technique) human reliability assessment method and co-developer of PREDICT (a Procedure to Review and Evaluate Dependencies in Complex Technologies). He is a Fellow of the Safety and Reliability Society, a founding member of the UK Human Factors in Reliability Group and a former Simon and Industrial Fellow at the University of Manchester. Mr Williams is a Contributor to a number of publications including, the "Guide to Reducing Human Error in Process Operation", the "Human Reliability Assessor's Guide", a "Guide to Task Analysis" and "Improving Compliance with Safety Procedures".
Email: jerry.williams@hse.gsi.gov.uk

FOREWORD

This volume is part of a long running scientific endeavour emerging from an international, interdisciplinary study group on "New Technologies and Work (NeTWork)". NeTWork is sponsored by the Werner Reimers Foundation (Bad Homburg, Germany) and the Maison des Sciences de l'Homme (Paris). The study group has set itself the task of scrutinising the most important problem domains posed by the introduction, spread and control of new technologies in work settings. Such a problem focus requires a strong multidisciplinary co-operation. For the last 20 years the group has organised a yearly workshop in Bad Homburg which tackles an emerging or problematic theme in this area. NeTWork operates with a small core group, which chooses the annual themes and plans and evaluates the workshops. Each workshop is co-ordinated by two or three "godfathers", who detail the theme, propose the prospective participants, structure the workshop and undertake, where possible, the production of a book based on the rewritten contributions, enriched by the discussions which take place in the three days. The participants are drawn from research, industry and government, in order to balance the insights from theory with the practical realities of work and organisation. Contributions are invited from both established and up-and-coming researchers. In its 20 years more than 200 persons from over 20 countries have participated, and this will be the twelfth book which has resulted from the workshops. One more is in preparation. In addition several of the workshops have resulted in special issues of relevant journals, such as Safety Science. It is always the aim of the books to be more than a collection of disparate contributions. The editors structure and order the chapters into a framework, in order to underline the advances and highlight the still remaining gaps and questions. In this way we hope that the books can serve as a reflection of the state of the art at the cutting edge of the subject. They are aimed at both researchers and at the advanced practitioners who have to struggle with the realities of managing or regulating the new technologies.

The original activities of NeTWork began with a wide coverage of themes, ranging from human error, training, information technology and distributed decision making. Recently its pre-occupation has been more specifically with a theme of great scientific and social significance: the safety of high technology systems and the role of the human and management contribution to their breakdown and control. This focus is justified by the public and scientific concern of the past two decades, following major disasters afflicting a wide range of complex technologies. These have struck nuclear power plants, ferries, aircraft, chemical installations, spacecraft, oil tankers, offshore platforms, trains, hazardous good transport and motorway pile-ups, tunnels, skyscrapers, telecommunications and complex computer systems such as the stock exchange. These "modern" disasters strike very different technologies, but share many factors in common; their occurrence in well-defended complex systems; the complexity of their causal roots; the dominance of human, rather than technical failures. They represent the visible tokens of the limits of our ability to control complex systems. It is these limits and the ways of both expanding them and yet staying securely inside them that the more recent workshops have explored.

This book, drawn from the eighteenth workshop in 2000, falls in this tradition, but also extends it somewhat by focussing on safety regulation. We look at how society tries to regulate the dangerous technologies and the companies which exploit them. For this book we, therefore, move up a system level from the topics which have occupied the NeTWork in previous years. Many regulators have taken part in the previous workshops and have contributed chapters to the books and journal issues, but their contributions have largely provided information on the pre-requisite conditions within which the technologies and companies had to operate. In this volume the regulator moves centre stage. We examine the dilemmas of the safety regulator, the tools and approaches available and the state of the art in making sense of and improving the effectiveness of the regulatory task.

As ever we wish to thank all the contributors to the workshop and to this volume for putting in so much work to rewrite their chapters on the basis of the discussions at the workshop and the reviewers comments. The framework into which we have tried to put their work, the first and last chapters of the book, is our attempt to make sense of their wise words. We hope it contributes to the understanding of how far this field of research and practice is and which questions still require study.

We would like to express our gratitude to the sponsors of the workshop, the Werner Reimers Foundation and the Maison des Sciences de l'Homme, for their long-term generosity. Thanks are also due to Bernhard Wilpert, chairman of the core group of NeTWork and to Babette Fahlbruch, its scientific secretary. Their tireless work ensures that the event is always impeccably organised and that the discussions flow as copiously as the wine. Thanks also to Wolter van Popta, who has ensured the layout and production of the book copy.

The editors: Andrew Hale, Barry Kirwan, Andrew Hopkins

1

ISSUES IN THE REGULATION OF SAFETY: SETTING THE SCENE

Andrew Hopkins & Andrew Hale

INTRODUCTION

In this chapter we set the scene for the book, which is based on the 2000 NeTWork (New Technology & Work) workshop. The chapter provides a framework in which we discuss the nature of safety regulation and sketch the range of dilemmas which it raises when we try to apply it in different activities, technologies and jurisdictions. This sketch will provide some guidance to the issues which the different chapters of the book will address. It also poses the questions which the participants in the workshop were asked to address in their papers, which have formed the basis of the chapters now presented here. The final chapter of the book returns to some of these questions and discusses what we have learned and how we might address them further.

Regulation is addressed primarily as an activity that operates externally to a technology, industry, transport system or other activity. Regulation is imposed upon the activity by parties outside, often the government on behalf of society, because there is concern or dissatisfaction with its effects. Regulation is therefore control or restriction. The motives for doing this can be sought in protection of weaker groups or interests in society, in prevention of the waste of resources (both natural and societal) and in guaranteeing a fair arena for market competition. This definition would seem to exclude one of the main themes of recent years, the concept of self-regulation, in which organisations control their own activities according to rules and criteria they set themselves. However, we see that simply as a regime imposed from outside, in which part of the regulatory process is delegated to the organisation, but under conditions which specify what must be regulated internally and how the outside world will check that this has taken place. Self-regulation fits, therefore, within our definition.

Regulation is one of the defining features of modern society. There are regulations governing taxation, pollution, traffic, trade, broadcasting, safety and so on, ad infinitum. It hasn't always been so. In simpler societies, the apparatus of the state was more rudimentary, and law coincided to a considerable extent with the deeply held beliefs of the community. But as the French sociologist Emile Durkheim (1917) pointed out a century ago, modern industrial society is highly differentiated, and legislators must enact laws designed to coordinate this diversity and to define and protect the public interest - as society, technology and the economy evolve. These laws do not necessarily correspond to deeply held sentiments. They are expedient - enacted to achieve public policy goals. They are arbitrary, in that they do not emerge in any automatic way from societal values and they may not be the only rules by which the public policy goals could be achieved. When we speak of regulation it is this type of legislation we normally have in mind. To give a trivial example, a rule requiring drivers to keep to the right side of the road produces orderly traffic movements, but so, too, does a keep to the left rule, and neither is based on any deep seated value.

What Is Distinctive About the Regulation of Safety?

In many areas of regulation it is possible to mandate the required outcome directly. For example, taxes must be paid, fishing quotas must not be exceeded and price fixing agreements must not be entered in to.

The regulation of safety cannot proceed in this manner. The policy objective is to prevent harm - harm to workers, to passengers, to local residents and so on. But the harm concerned is almost never intentional, at least not from the point of view of the local participants; it is in some sense almost always accidental. From a statistical point of view it may be highly predictable that a number of deaths or injuries will occur in a given activity or industry, but not to whom or exactly from what. Stone (1975) defines behaviour which results in this sort of harm as 'provisionally undesirable', as opposed to the absolute undesirability of such crimes as murder. This distinction also implies that society gains an advantage from the 'provisionally undesirable' activity, which is impossible to separate from the risk that goes with it. Because of this it is not possible to enact regulations which prohibit the harm or to impose an absolute duty on those who operate hazardous technologies that they do no harm. At most we can require the adoption of certain procedures which are designed to protect against harm. Safety regulation is necessarily indirect in this way. It is only direct when we enact provisions requiring parties to repair or pay for the results of accidents for which they are deemed responsible.

Safety regulation is about the regulation of risk. In this respect it differs fundamentally from many other kinds of regulation. The risk of harm is ever present as we go about our lives - at work, as we travel, as we consume. All that can reasonably be required of those who control our workplaces, who transport us or who make the goods and services we consume is that they *minimise* the risks to which they subject us. The German sociologist Ulrich Beck (1992) argues that we face unprecedented environmental and technological risks and that a concern with risk - with quantifying it, distributing it, avoiding it, controlling it - is what characterises contemporary society. We live, he says, in a "risk society". Leaving aside Beck's hyperbole, it is clear that the regulation of risk is now one of the central regulatory tasks.

Is Regulation of Safety Necessary At All?

A fundamental question is whether the external regulation of safety is necessary at all. Some economists have argued that safety is simply good business and that accident prevention can ultimately be left to the market (Oi, 1980). They point out that accidents are costly, in terms of lost production and destruction of assets. Moreover the damage claims brought by injured parties under the ordinary law of torts or equivalent claims regulations in other systems, provided that they are fair and effective, can provide a strong economic incentive to minimise the risk of harm to individuals. Enlightened self-interest, even in the absence of any specific safety legislation, should therefore serve to minimise risk.

There are a myriad objections to this view which will not be rehearsed here (see e.g. Robinson 1991, Noble 1986). Not the least of them is that the damage claims system is already a form of regulation in which the victim (the weak element in the system) is enabled to gain recompense from the party that has caused the damage. The question is raised at this point, however, because it provides a context for two of the chapters in this volume. The chapter by Baram demonstrates just how risky an area of activity can become, if it is left to the market to regulate. His chapter provides an empirical demonstration of the need for regulation. De Mol argues in his chapter that it is true that civil litigation can promote safety. However, he shows that there are numerous problems with reliance on this mechanism and that it can really only function as a backup, where the regulatory system has failed to control the problem in a more proactive way.

What Is Required of the Regulated?

The fact that harm cannot be simply prohibited is the source of many of the distinctive dilemmas of safety regulation. If the regulated cannot be required by law to do no harm, what *should* they be required to do? What is the role of the regulators in these circumstances? Should they be seeking to punish or persuade? These and other issues will be addressed in the following sections.

Prescriptive regulation

The traditional regulatory strategy which emerged in response to the industrial revolution was to require those who create the risks, in particular employers, to abide by very specific technical requirements, for example in relation to ladder heights, door sizes, electrical safety, scaffolding requirements etc. As industry developed and technology advanced the volume of such prescriptive legislation steadily increased and by the third quarter of the 20th century, according to critics it had become unwieldy, unenforceable and unworkable (Hale 1978, Robens 1972). Amalberti (2000) chapter demonstrates some of the problems associated with the prescriptive approach to air safety and he argues that an increase in volume of prescriptive rules in recent years has not led to an increase in air safety.

Goal- oriented legislation

The 1972 Robens report in the UK and earlier Scandinavian legislation marked the beginning of a revolution in the general approach to regulation of safety. Robens recommended that, rather than specifying standards and procedures to enhance safety, legislation should specify the policy goal itself – safety, and require employers to ensure the safety of their workers, so far as reasonably practicable. In principle it would be up to the employer to decide how best to achieve this goal. Many jurisdictions have now adopted the principle of goal-setting legislation, also known as outcome-oriented or performance-based legislation. This changeover provides the theme, or the backdrop, to many of the chapters in this book.

The new approach has, however, been bedevilled by the problem identified above: although the goal was safety, employers could not be expected to guarantee the safety of their workers absolutely. The requirement, therefore, was to ensure safety so far as reasonably practicable. But how could employers be sure that they had done what was reasonably practicable? They needed guidance, in the form codes of practice and guidelines. Provided they complied with these inspectors would find no fault when they visited the organisation and, in the event of an injury or fatality occurring, courts would conclude that it had done what was practicable and was therefore in compliance. In practice, therefore, the regulatory regime tended to revert to prescription, a more flexible form of prescription to be sure, but prescription nevertheless (see further, Hopkins, 1994).

Safety management systems and the safety case approach

In recent years regulatory requirements have undergone a further shift. Increasingly legislation is requiring employers and facility operators to introduce safety management systems (Gunningham and Johnstone, 1999). In some jurisdictions this has simply been a way of specifying what it is that employers must do to comply with the general obligation to ensure safety. But in the case of major hazard facilities, safety management systems have been mandated as part of a broader, safety case strategy.

Following major accidents in England (Flixborough, 1974) and Italy (Seveso, 1976), the EU issued a directive to its member states to implement a safety case strategy for major hazard facilities (the 1982 Seveso directive, replaced in 1996 by the Seveso II directive, see Oh's chapter in this volume). The radically new feature of the safety case strategy was that it required operators of major hazard facilities to *demonstrate* to the regulator that they had identified, assessed and controlled the hazards in question. Among other things the Seveso directive spells out in some detail the nature of the safety management systems which operators must implement. Here then is a further reversion to quite detailed prescription. The difference is that whereas earlier regimes specified technical details, the most recent approaches are specifying details of how safety is to be managed. Williams' chapter provides a good illustration of this approach, in his description of the regulation of the nuclear industry in the UK.

Some safety case legislation (such as the regime in Victorian major hazard plants described in Hopkins' chapter) treats the safety case more or less as a one-off, static regime. The strategy requires a concerted effort to identify, assess and control hazards at the outset, but once this has been done and the case has been accepted by the regulatory authority, there is a temptation for facility operators to rest on their laurels. However, studies of high reliability organisations show that the key to safety is an on-going mindfulness about the possibility of failure and a determination to identify early warning signs of trouble and to learn from such incidents (Weick, 1999). Safe organisations are those which have institutionalised procedures for identifying and responding to critical incidents and for learning from them. They are learning organisations (Senge 1990), which constantly attend to their safety management system and its performance in risk control. They adapt it to new insights, new hazards, new technology and new expectations from society, government or market. Rasmussen (2000) emphasised the point in a paper presented to the workshop, but not included in this book, called: "safety - a communication problem?"

Understanding that the problem is one of mindfulness does not however undermine the safety case approach; it strengthens it by highlighting aspects of the safety management system which need to be stressed if organisations are to be reliably safe. Management systems need to include procedures for responding to critical incidents and learning from them. This is what Bavarian nuclear power stations, for instance, have done, as Walther's chapter shows (see also Becker 1999). Hopkins' chapter in this volume describes the way the regulation of coal mining in Queensland has also moved in this direction by requiring companies to define triggers which will set parts of their safety management system in motion. The regulation of food safety is also based on this model of identifying and responding to critical incidents (HACCP – Hazard Analysis and Critical Control Point; see NACMCF 1997). Becker's chapter also contains the suggestion that regulators monitor the capacity of organisations to attend to warning signs and to learn from their own and others' mistakes. In the absence of regulatory requirements to this effect there is no guarantee that they will do so.

Requiring that organisations and individuals report potentially incriminating incidents is of course problematic (Baram 1997) and any such regulatory requirement would need to include protection for reporters. But the incident reporting systems in operation in the airline industry demonstrate that the problems are not insuperable (Reason, 1997:196-205).

A Framework for Regulation, Rule Making & Enforcement

The preceding discussion has identified three models for the content of safety regulation - technical prescription, goal-oriented legislation, and legislation requiring safety management systems – and it implicitly assumes that these models represent an evolutionary sequence. This was indeed the case for the UK and countries which follow the UK system, such as Australia. But this sequence was not universal. In some areas, prescriptive technical regulation remains the preferred strategy and in others, as Hovden's chapter shows, elements of technical prescription are being reintroduced in what for some time has been goal-oriented legislation.

We can picture these different models in the form of a framework (figure 1), which depicts three different levels of intervention in risk control.

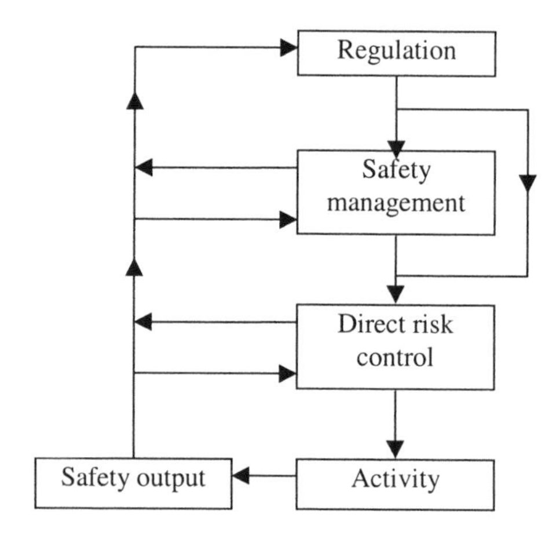

Figure 1: Three levels of risk control (based on Hale et al 1991)

The technical prescription regime specifies rules for the direct risk control function. The goal-oriented regime tries to specify the safety output, but not the rules to achieve it. The safety case/ safety management regime specifies the rules for the safety management block and/or requires that block to demonstrate how it structures and runs the direct control activities. The feedback arrows on the left of the figure show that the regulator has three sources of information available about the system's performance: the safety output, the structure and functioning of the direct risk controls and the structure and functioning of the safety management system. A regulatory regime can choose to place its emphasis differently in both what it specifies as rules and what it looks at as evidence of compliance with the rules, or quality of performance. The various chapters in the book illustrate these different choices and the regulatory dilemmas which form the main content of this book. In the next section we present a further development of this model in order to provide a structure for these dilemmas.

We have indicated above that regulation is about rule making and ensuring that parties comply with the rules. We can represent this process as one of problem solving – see figure 2 (Hale et al 1997). Indeed all three levels of direct risk control, safety management and regulation can be usefully seen in this way.

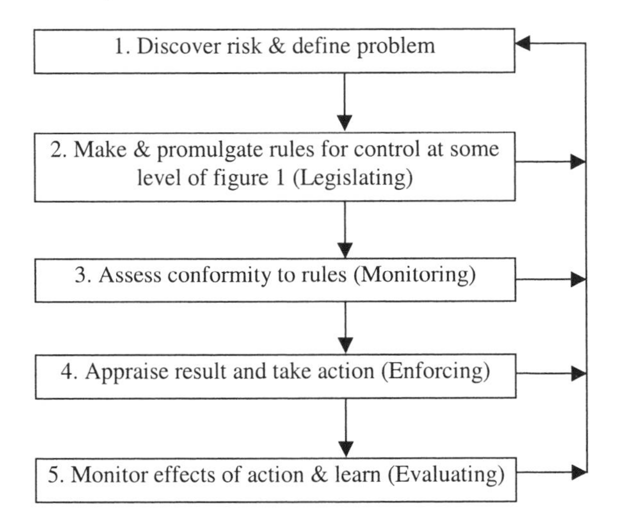

Figure 2: Regulation as problem solving

This figure matches the framework proposed by Eijlander (2001) for considering the dilemmas and choices of regulation in any given domain. The terms he uses are placed in brackets in the steps of the figure; only step 1 is missing in his formulation. Figure 2 divides this process of regulation into

5 steps along these lines. These steps will be used below as framework to discuss the essential tasks of the regulator or regulators and the choices which have to be made in carrying them out. As we shall see, there is often another overarching task of the regulator, particularly if we are talking about the government regulator. This is to communicate to the public and other interested parties about the process of risk control, how it operates and how successful it is. This is an essential role in all cases where the regulator has been delegated the task of looking after society or its weaker elements in the face of the risks of an activity. Reiman & Norros' chapter discusses this role explicitly, but other chapters also touch upon it and its potential conflict with the monitoring and rule making tasks.

In the final chapter we return to each of the five steps of figure 2, in order to summarise some of the lessons from the workshop. In the final part of this introductory chapter we raise some more general issues and questions which the workshop addressed.

REGULATION BY NON-GOVERNMENTAL BODIES: WHO DOES WHAT?

It may have seemed up to now that we have equated "regulator" mostly with a government agency and with the process of law making in a country. In this respect we have often thought of the government ministry as the rule maker and the inspectorate as the compliance tester and rule enforcer. This is a traditional equation, and government still has a strong role in safety regulation in most countries. However, in parallel with the shift of regulatory paradigm from technical prescription to safety management has gone a shift from centralised government regulation to a more decentralised system in which far more actors are involved. This is tantamount to the allocation of some or all of the tasks in figure 2 to non-government bodies. We briefly review here the other potential actors in regulation and will return to them in discussing the steps in figure 2.

Compensation system

The common law system of workers compensation developed in countries like the US and the UK can also be seen a part of the non-governmental regulatory apparatus. Over time judges have developed a series of rules governing the circumstances under which compensation would be paid. Such systems serve in an obvious way to regulate the compensation paid to workers. They are also widely assumed to encourage employers to attend to safety, but we have little evidence to indicate whether such an effect does occur. Even where common law systems have given way to state workers compensation systems with premiums set to reflect the claims experience of employers, we have little evidence whether this has contributed to safety (however see Larsson & Clayton 1994). On the other hand, in many European countries social insurers responsible for paying workers compensation have developed and imposed safety rules as a condition for insuring the workforce. In these circumstances the insurance system has had a pronounced regulatory effect. Private insurers have followed suit in relation to other damage, such as fire, etc.

Standards

Technical rules have always been potentially open to being written, or at least influenced by professional and scientific bodies such as the societies of engineers, chemists or hygienists, as well as by insurers (both public and private). In the last years, however, there has been increasing use of the system of national and international standardisation. National and international non-governmental organisations, with input from relevant industries, consumers and employee representatives, have developed a variety of standards covering safety. Manufacturers and industry bodies have been added to the list of actors writing rules adopted by the regulatory system. These standards often include the kind of prescriptive detail which governments no longer feel it appropriate to include as formal regulatory requirements. Nevertheless, government goal-directed legislation will sometimes make reference to these standards. Moreover, because the standards have industry agreement, firms which do not comply will be judged not to have done what is reasonably practicable, or what is seen as 'state of the art', should the matter end up in court. In this sense these standards are enforceable by governments. In this way government and non-government regulatory efforts reinforce each other.

Private sector certification

It is not only government inspectorates who police the rules. A variety of non-governmental, third party bodies are available to certify whether an organisation is in compliance with relevant standards, as Gundlach describes in his chapter. But why should an organisation subject itself to this scrutiny? The reason is that many businesses will now only do business with others whose products or services have been certified by some third party as complying with the relevant standard. An example is the petrochemical industry and its on-site contractors carrying out maintenance or construction work (e.g. SSVV 1997). Here, therefore is an area of non-governmental policing which is entirely market driven and which is having an increasing influence on regulation. Governments may latch onto this private initiative by requiring certification by law (of competence, or management systems), thus combining public with private regulation.

Intra-organisation rule making

Safety within any organisation depends on people following safe operating procedures. As governments retreat from responsibility for specifying such rules and instead enact goal-directed legislation, it is left to organisations themselves to develop their own detailed rules, as Hale et al point out in their chapter. Again therefore there is an interaction between government and non-government sectors. Governments require that rules be made but organisations themselves make the rules to which they will adhere. Professionals within organisations have always operated in this way, as de Mol describes in his chapter on medical devices. The issue in all these cases is whether the organisations and professionals can be trusted to self-regulate.

Product stewardship

Because of its vulnerability to public opinion, the chemical industry developed in the early 1990s a system of industry self-regulation known as responsible care (Gunningham, 1995). One aspect of this was the concept of product stewardship. The idea was that the manufacturer would oversee the use of its product after it left the factory gate, requiring all those who transported it, sold it or finally used it to comply with certain standards. Here again is a system of regulation operating entirely within the private sector. Governments can, however, encourage the spread of product stewardship. Suppose inspectors collected the brand names of products involved in accidents they investigated. They might then at the end of the year note that there had been, let us say, five accidents involving roll-overs by "Anderson" tractors. The regulator could then approach the company and suggest that it had an interest, not only in tractor design, but also in how its tractors were being driven by end users. If the authorities were able to calculate rates and compare the accident rates of different brand name tractors, the leverage could be increased. Governments are sorely in need of finding ways to promote safety in the context of small business and it is precisely in this context that product stewardship might be most effectively promoted. It is this sort of leverage which was used successfully by Ralph Nader in his campaigns to force the US automobile industry to increase safety.

The last years have also seen the increasing influence of other pressure groups of victims, consumers or environmental interest groups in applying pressure to organisations and achieving changes which the traditional regulatory system had failed to make.

All of these groups, and conceivably others, can play a role in regulation as we have defined it. The allocation of tasks between all of these actors is, therefore, a choice which a regulatory regime can make.

Questions for the Book

Having provided a framework for the concept of regulation we can now pose a number of questions. These were, in essence, the questions which the contributors to the workshop were asked to address in their papers, which have now been collected as the chapters of the book. They were as follows:

1. What are the various forms of regulatory control for safety which have been applied and how successful are they? How are the five steps of figure 2 implemented in practice in different technologies, activities and jurisdictions? What dilemmas and advantages accrue from the different regulatory regimes?
2. How are the regulatory regimes matched to the risk or activity to be controlled? Rasmussen & Svedung (2000) have made a strong case that the direct and safety management risk controls need to be closely matched to the activity and risk in question. Is this also true of regulatory actions? If so, how specific must the match be?
3. As a corollary to question 2, what can we learn from one regime and apply in another technology, risk, activity or jurisdiction? What characteristics of the regulatory approach, or of the risk or activity to be regulated determine whether a regulatory approach is portable or not? How do we apply these lessons to new technologies which have so far not been regulated?
4. Do regulatory regimes go through some necessary process of evolution or revolution to reach the desired level of control? How do we know when to change our regulatory paradigm for one which has a chance of being more successful?

We do not claim that these are all the relevant questions to be asked about regulation. Nor can one workshop or book hope to answer even all of these questions. The chapters presented here attempt to illustrate a range of experience, which describe the state of the art on answering them in the jurisdictions concerned. The last chapter of the book will return to them and take stock of what we have learned from the contributions. It will also indicate some gaps in the coverage of the book and try to point the way to future research and application.

REFERENCES

Amalberti, R. (2000). The paradoxes of almost totally safe transportation systems. *Safety Science* **27** (2/3) 109-126

Baram, M. (1997). Shame, blame and liability. In A. Hale, B. Wilpert, & M. Freitag (Eds.), *After the Event: From Accident to Organisational Learning*

Bardach, E., & Kagan, R. (1982). *Going by the Book: the Problem of Regulatory Unreasonableness*. Philadelphia: Temple University Press.

Beck, U. (1992). *Risk Society*. London: Sage.

Becker, G. (1999). From theory to practice - on the difficulties of improving human-factors learning from events in an inhospitable environment. In J. Mismuni, B. Wilpert, & R. Miller (Eds.), *Nuclear Safety: A Human Factors Perspective* (pp. 113-125). London: Taylor & Francis.

Durkheim, E. (1993). *The Division of Labour in Society*. New York: Macmillan.

Gunningham, N. (1995). Environment, Self-Regulation and the Chemical Industry: Assessing Responsible Care. *Law and Policy*, **17**(1), 57-109.

Gunningham, N., & Johnstone, R. (1999). *Regulating Workplace Safety: Systems and Sanctions.* Oxford: OUP.

Hale, A.R. 1978. *The role of HM Inspectors of Factories with particular reference to their training.* PhD Thesis. University of Aston in Birmingham. UK.

Hale A.R., Goossens L.H.J. & Oortman Gerlings P. 1991. Safety management systems: a model and some applications. Paper to 9th NetWork workshop on Safety Policy. Bad Homburg. Hale A.R., Heming B. Carthey J., & Kirwan B. 1997. Modelling of safety management systems. *Safety Science* **26** (1/2) 121-140.

Hopkins, A. (1994). Compliance with what? The fundamental regulatory question. *British Journal of Criminology*, **34**(4), 431-443.

Larsson T.J. & Clayton A. (eds.) 1994. *Insurance and prevention: some thoughts on social engineering in relation to externally caused injury and disease.* IPSO. Stockholm.

NACMCF (1997). *Hazard Analysis and Critical Control Point Principles and Application Guidelines* No. National Advisory Committee on Microbiological Criteria for Foods.

Noble, C. (1986). *Liberalism at Work: The Rise and Fall of OSHA.* Philadelphia: Temple University Press.

Oi, W. (1980). On the economics of industrial safety. In D. Petersen & J. Goodale (Eds.), *Readings in Accident Prevention* (pp. 65-91). New York: McGraw-Hill.

Rasmussen, J. (2000). Safety – a communication problem? Paper to the NeTWork Workshop on Safety regulation. Bad Homburg

Rasmussen J & Svedung I. 2000. *Proactive risk management in a dynamic society.* Swedish Rescue Services Agency. Karlstad, Sweden.

Reason, J. (1997). *Managing the Risks of Organisational Accidents.* Aldershot: Ashgate.

Robens. Lord. (1972). Safety and health at work: report of the committee. HMSO. London.

Robinson, J. C. (1991). *Toxics and Toil: Workplace Struggles and Political Strategies for Occupational Health.* Berkeley: University of California Press.

Senge P.M. (1990). The Fifth Discipline: the art and practice of the learning organisation. Doubleday. New York.

SSVV (1997). VCA: Safety checklist for contractors. Leidschendam. SSVV.

Stone (1975)

Weick, K., Sutcliffe, K., & Obstfeld (1999). Organising for high reliability: processes of collective mindfulness. *Research in Organisational Behaviour*, **21**, 81-123.

PART I: THE CHANGING FACE OF REGULATION

The main section of the book presents a series of studies and descriptions of the way in which regulation is currently carried out in several industries and countries. The contributions span six countries (Australia, Netherlands, UK, Germany, Norway, Finland) and six industries (nuclear, rail, offshore, coal, chemical, and the general small and medium sized enterprise (SME)). Four chapters are by practitioners employed now or previously by the regulator, six are by researchers who have studied either the regulators themselves, or the results of their work in the form of accidents or technical and organisational changes. Although these ten chapters cannot cover all forms of regulation, they raise a wide range of issues about what works here.

This chapter, by Larsson, is the first of two contributions to this volume from Australia. A Swede who has been in Australia for more than a decade, Larsson writes with a truly international perspective. This contribution concerns the problems of regulating health and safety in small business and draws particularly on Larsson's experience in the state of Victoria. But the insouciance of small businesses in relation to workplace hazards and the difficulty which regulators have in reaching small business are world wide problems and readers will no doubt be aware of similar problems in their own environment.

Larsson makes the important point that economic activity is increasingly outsourced, decentralised and fragmented and that large and increasing proportions of workers find themselves in small business environments. The injuries which befall them in these circumstances are largely hidden from the wider society - hence his title: "the pulverisation of risk and the privatisation of trauma". He is the only contributor to this volume to confront this issue explicitly; nearly all the other contributions concern large organisations where failures impact on the public at large as well as on the workers.
Larsson also identifies the role of economic forces in creating health and safety problems. This is a theme which nearly every contributor touches on.

A particular problem which Larsson describes is the commercialisation of the occupational health and safety regulator and its transformation from an organisation which sees the workforce (and the public) as its client to one which views the employer as client. This has been a major issue in Australia and readers may wish to consider to what extent this issue has arisen in their own context.

2

New Technologies and Work

Pulverization of Risk - Privatization of Trauma?

Tore J. Larsson

1 Legal Requirements and OH&S

The development of OH&S in the advanced export-oriented industrial economies has shown that technological development, investment and innovation are critical factors in the reduction of occupational injury and disease. Competitive industrial development has, in most cases, been underpinned by a regulatory framework demanding equal compliance from all in the market, combined with reasonable levels of inspectorate detection. The regulatory framework has, apart from protecting workers against trauma and disease, served to prevent unlawful shortcuts and undeserved market advantages.

The involvement of workers in quality improvement and innovative development of products and tasks represents another critical ingredient to the improvement of the working environment and to worksite climate, and this, in turn, has been of importance to OH&S in the large industrial corporations. Examples from Japan, Germany, USA and Scandinavia bear witness to this (Oxenburgh, 1991).

The successful elimination of extreme industrial hazards (eg. modification of chainsaws, redesign of controls for power presses, roll-over protection for tractors) are examples of selective, demand-driven regulatory approaches which have been instrumental in reducing fatalities and the incidence

of severe trauma Pettersson, 1982; Mäkinen, 1982; Springfeldt, 1996). The benefit of such successes have been substantial.
The safety inspectorates of the Western industrial world changed their main paradigm in the 1970s. From the Norwegian and British experiences in the off-shore oil and gas fields came the notion that Government could not compete with the competence of the high-tech process industry and thus should instead concentrate on ensuring that systems of control were in place.
This represented a big step away from the specialised technical inspector, implied a judicial or organisational audit and was easy to combine with a general principle of "duty of care" (Lord Robens, 1972).
The efficiency crisis of the industrial safety inspectorate in the 1990s has been evident throughout the modern industrial world. In countries, where the paradigm has shifted from "prescription" to "performance", the inspectorate has seen reduced resources, reduced pro-active inspection and control activities and the emergence of rather unconvincing, but cheaper, forms of advisory services. - Todays big industrial company has little need for the inspectorate. Large numbers of medium-size companies often have great demand for OH&S advice and services, but they will not take their problems to the inspectorate. Small companies and the self-employed mostly exist outside control - and often outside the law (Mayhew, 1997; Mayhew et al, 1997; Johnstone, 1994).
The contradiction between inspection and control, on the one hand, and prevention and industrial development, on the other, has become more pronounced.

Compliance control - mixed and contradictory roles

The following observations have been made in relation to the State of Victoria, Australia, but represent an approximate picture of the development in other Australian jurisdictions.

Given the change of focus, Labour Inspectorates (or similar but euphemistically re-named organisations) have recruited more staff with specific advisory or consulting competence - not necessarily with the skills and inclination to enforce.
Consistent with the obligation to ensure that compliance remains in the workplace (and not with Government) new performance based standards require hazard identification, risk assessment and risk control. Such activities are basic and necessary to a functional system of safety management, but the requirement introduces a subjective element into the process which continues to cause problems and disputes as to whether what has been done complies with the legal requirement (Dell, 2001).
The level of enforcement is low. Field staff, and industry, grapple with the concepts in a mixed facilitation/enforcement environment and consequently find it difficult to clearly explain and define the requirements in workplaces. Inspectors try to find the 'firm ground' in these standards and as a result find it difficult to enforce - particularly with the mixed messages they get from their organisation in terms of their core role.

In spite of the change of terminology, there is a reluctance on the part of employers to seek advice and assistance from the inspectorate. Information networks have been established to provide "field officers" (who are not inspectors) with support to deliver this service in a non-threatening way. However, although such an information network can provide a potentially valuable service, it is still part of an enforcement organisation. The fact that companies still perceive a threat from their involvement remains unresolved (Larsson, 1998).
The combined adviser/enforcer role allows for the exercise of considerable discretion on the part of the inspector with a corresponding potential for corruption. There is obvious concern on the part of the client that discretion may be exercised capriciously or arbitrarily - the industrial company can have no confidence in the evenhanded application of the law - at the end of the day it may depend on whether or not the inspector likes them or how obsequious they were to the inspector.
The advisory, training, and information roles take inspectorial staff away from what is their compliance control function - currently less than 30% of available time in most inspectorial systems is spent in actually undertaking workplace visits. As a result, most employers have no expectation that they will be visited by an inspector (Larsson, 1998).

Industrial relations?

To have the OH&S regulatory and enforcement agency clearly identified and associated with the issues of employment and industrial relations puts it in the category of contractual, employment and civil law rather than criminal law. This detracts from the impact and influence of the OH&S legislation and confuses the community as to where these obligations fit.
Industrial relations and OH&S will, from time to time, impact upon each other in relation to particular issues, but to portray them as fundamentally linked will serve to keep OH&S firmly inside the factory gates and out of the mainstream of legal obligations, duties and moral responsibility in society.
There is nothing wrong with the promotion of good OH&S for improved business performance, but as a primary objective of the industrial safety agency it could often, in fact, be in conflict with the enforcement of OH&S law, which mainly seeks to protect the health and safety of employees and the environment.
The creation of audit functions and certification processes (eg. SafetyMAP, FiveStar, etc) has arisen from the desire to further undertake roles other than inspectorial (safety auditors are not authorised as inspectors). In terms of the encouragement of industry into the certification process, the Inspectorate as the regulatory agency is not able to provide any real positive incentives for organisations to take this up; the Inspectorate can only promise to stay away if companies are certificated and clearly, in a number of circumstances, the Inspectorate cannot stay away and still fulfil its role as the administration agency of the legislation.

Industrial lawlessness

The present industrial safety practices, particularly in the cultural sphere dominated by the British legal tradition (viz. UK, Australia), are covered by regulatory frameworks designed not to instruct or şet minimum requirements, but to be used for prosecution when a worker has been severely injured or killed.
The pro-active control of industrial practices has been cut back to such a level that the perceived risk of violations being detected is minimal. For all practical purposes industry has been left to self-regulation; a system of informal and uncontrolled cross-subsidization, where good performers pay a higher price per produced unit than poor performing competitors, who can undercut and gain market advantages by disregarding OH&S in an environment of deregulated safety.
The role of the industrial safety inspectorate has been redefined and couched in such terms as to imply that the main activities for the inspectorate are not to control and monitor the law and order of the working environment and the protection of individual workers' health and safety. Instead their aim is "to provide positive cultural change", "to simplify the OHS regulatory system", "to improve services to our customers", "to develop the network of OHS service providers", "to pro-actively contribute to the improvement of OHS through leadership, innovation and service excellence", and (at the bottom of the list) "to reduce the number of fatalities ...particularly in the Agricultural sector" (HSO Victoria Business Plan 1995/96; pp 18-21).
The police force of the work environment have decided to stop the pro-active enforcement of the law, to take off their uniforms and become consultants and advisers to an unknown number of reluctant and suspicious industrial customers.

2 SMALL BUSINESS

While political trends in many countries have led to "smaller" government, reduced public service, deregulation and privatisation of what used to be government responsibilities, there has been increased attention focused on small and medium-size enterprises (SME).
Economies and labour markets, which until the 1970's were dominated by large employers, have seen multi-nationals down-size, out-source and sub-contract; big conglomerates have shed non-core activities and forced local labour markets to restructure. Average unemployment figures have increased and have remained at increased levels.
The proportion of the total UK labour force employed by companies with less than 50 staff increased from 33% in 1979 to 43% in 1986 (Quinn, 1997). The US economy shed 639,000 full-time jobs between 1989 and 1992, while at the same time there was an increase of 894,000 part-time jobs (Morrisey et al, 1994).
Increasing proportions of the work-force in many industrial countries have become marginal, or more loosely attached, to the labour market. Full-time employment in large public and private corporations has decreased. Partial or temporary unemployment, partial or temporary self-

employment, and small family-based enterprises have become more common. The size of the informal, untaxed and uncontrolled economic sector is considerable and probably increasing. These trends are clearly discernible in most modern economies.
With the increasing fragmenting of economic activity comes naturally the pulverisation of risk. Active risk management in the large organisation has often implied the shedding of especially hazardous tasks onto external and mostly small sub-contractors. The issues of work-related traumatic injury, workers' compensation and safety at work in relation to small business have become increasingly relevant and also more widely reported (VWA Advisory Committee, 1994; Stanley et al, 1996; Mayhew, 1997; Mayhew et al, 1997).
In some sectors (eg. chemical, mining, construction) major conglomerates have retained stewardship over sub-contractors and require that safety management practices are applied, but in many areas of sub-contracting and out-sourcing, safety is not managed well and the consequences of poorly controlled OH&S will be picked up by the injured worker or be shunted onto the public purse (Larsson & Betts, 1996; Mayhew et al, 1997).

Victoria, Australia

Between 1983/94 and 1994/95 the number of small businesses in Australia increased by 43% or 3.3% per annum. 710,000 Australians in 1983 worked in their own business. 12 years later this had increased by 33%; in 1994/95 950,000 Australians were employed in their own business (ABS, 1996).
291,000 non-employing businesses (partnerships and sole traders) in Australia in 1983 grew to 428,000 in 1994/95, representing an increase of 47% in twelve years. During the same period the number of businesses employing less than 10 people increased from 224,000 in 1983 to 315,000 in 1994/95, an increase of 40% (ABS, 1996).
The Victorian WorkCover Authority data for 1992/93 showed that, based on reported payroll, more than half of all Victorian employers had a remuneration of $50,000.- or less, which represents around two full-time employees; three-quarters of all employers have a remuneration of less than $125,000.-, or an estimated maximum of 5 full-time employees (VWA Advisory Committee, 1994).
Whereas the sole traders, partnerships and purely self-employed are not required to insure under Workcover, in reality a large proportion, perhaps the largest, of Victorian business establishments presently registered with Workcover are 1 - 2 people, self-employed, partnerships or family operations, who have been advised to incorporate.
Definitions of small business vary but often refer to number of employees, ie. normally <20 employees, or in manufacturing <100 employees, often exclude agricultural businesses, often refer to limited turnover, ie. <$5M or <$10M. The Australian Bureau of Statistics also defines as small business a company independently owned and operated, but closely controlled by the owner/manager, who also contributes most of the operating capital and makes all the important decisions in the business (ABS, 1996).

ABS estimated that there were 786,000 small private sector businesses (excluding agriculture) operating in Australia in 1994-95, employing some 2.7 million people (ABS, 1996). Small business account for more than 96% of the total number of all businesses in all States and Territories, with the single exception of the Northern Territory, where 93% of all businesses were classified as small. Victoria has some 193,000 small businesses employing more than 680,000 persons, or just over 50% of the total workforce in the State (Larsson & Betts, 1996).

Safety in Australian small business

In a report to the Victorian Workcover Advisory Committee on "Incentives for Accident Prevention Among Small and Medium Sized Employers" (VWA Advisory Committee, 1994), the authors concluded that SMEs had

- lower claims incidence than large employers,
- higher average claims costs than large employers.

The authors attribute the higher average claims cost among small businesses to poorer "return-to-work" options and a lack of information about prevention and risk management, which would in turn generate more serious and costly injuries. In commenting upon the positive correlation between size and claims incidence, apart from accepting the logic of increased exposure of more staff to hazards resulting in more injuries, the authors raise the issue of claims suppression (VWA, 1996: 17-18).

Potentially, several aspects crucial to decision making in small business might help explain the high average claims costs. It is almost certain that there are structurally and size-biased exposure to hazards among some types of industry; some high-risk activities are always done by independent (small) operators.

There might be under-reporting of minor-medium severity injuries due to the inability of the owner/manager to take time off in a small business; eg. studies indicate that farmers only report serious injuries (Larsson, 1990b). There might be potential claims suppression due to perceptions on the experience rating aspects of the premium system. Empirical material on the qualitative aspects of small business decision-making in relation to safety is scarce.

However, two extensive reports recently published by the Queensland government in co-operation with Worksafe Australia, provide a wealth of information on the risk exposure, knowledge of and attitudes to occupational hazards among repair garages, cafes and restaurants, newsagents, building contractors, cabinetmakers and shopfitters, and demolishers (Mayhew, 1997; Mayhew et al, 1997). The aim of these studies has been to assess what barriers exist to implementation of solutions to OH&S problems, and which are the optimal ways to communicate OH&S to small business operators. The authors' main conclusions are that

- OH&S messages have failed to penetrate the very small businesses because the methods used are inappropriate, ie. based on those employed with large business,

- the very small business owner/manager is generally unaware of the OH&S legislative framework; they know, use or understand very little about OH&S,
- local, face-to-face, industry-specific OH&S advice seems to be a viable way to improve OH&S performance among small business.

(Mayhew, 1997)

In view of the fact that there seems to be, at least in Victoria, a strong overlap between the really small (self-employed) family business and the typical small enterprise registered with the Workcover Authority, some aspects crucial to decision-making in such businesses need to be considered:

- Ethical decisions among small business owners have been shown to be more dependent upon the private sphere, family and individual beliefs than they are among professional managers (Quinn, 1997).
- Entrepreneurs from smaller firms tend to be less "comprehensive" in their decision behavior, ie. less likely to follow a formal rational decision process, than professional managers from larger firms (Smith et al, 1988).
- Family business owners tend to refrain from drawing money from the business, they are less career-oriented, more centralised in their decision-making, and less likely to seek external advice than professional managers (Dyer, 1986; Leach, 1991; Cromie et al, 1995).
- In family firms the assignment of tasks, the grouping of work activities, the flow of tasks and information, and the control of work processes are informally organised (Whisler, 1988).
- Owners of small firms tend to interact with and influence their staff directly, whereas professional managers, who are required to justify their actions to boards, shareholders, and external organisations, initiate numerous formal reporting procedures to cope with their responsibilities (Goffee & Scase, 1985).

There is no simple relation between high-risk occupational activities and the size of the enterprise. Much of what is written and discussed about OH&S in small business implies that the lack of safety officers, formal structures for company administration, the lack of formal procedures for HR and OH&S management, and the lack of formalised industrial relations constitute barriers to improved work environment and workers' health. However, there is not much empirical or scientific proof for this and it might be more a reflection of frustrated communications efforts than a proper analysis of the true state of OH&S in small business.

Since the 1970s change of paradigm in government control of industrial safety, the "duty of care" compliance control approach has been specifically tailored for the OH&S administrative resources of the large industrial corporations (Gun, 1992; Larsson, 1997).

Out-sourcing, sub-contracting and self-insurance among large corporations in Australia has increased, and as growing proportions of the labour force are self-employed, in small business or have been forced into informal or temporary employment, the compliance control arm of government has been left with inadequate tools to do the job effectively.

In an interesting comment to the problems of OH&S promotion and risk communication, Holmes (1994:30-31) discusses the different perspectives on risk, which will define the way priorities are set and how the task of prevention is perceived. She argues that technical risk assessment, cognitive approaches to risk and social/cultural theories on risk will generate very different communication and prevention strategies in OH&S.

Decision making and occupational health

Most reasoning around incentives for occupational injury prevention and improved work environment assumes that there are certain functional economic incentives, combined with legal requirements, which if presented to an employer in a suitable way will result in decisions on investment on the part of the employer so that risk exposure is decreased and injury rates lowered. This rests on the view that conditions creating occupational hazards, and successfully applied preventative measures, are, to some extent, information or decision-making problems for the employer. Further conditions specific to the occupational injury problem are that

- the problem is perceived as related to possible employer activities such as investments; ie. perceived as possible to be influenced by the employer,
- the solution to the problem is at hand, possible to access by the employer in a reasonably straightforward way, and
- the investment to prevent injury is comparable to the perceived prevention-induced gain over a reasonable period of time.

These conditions are not always present. To invest in the prevention of occupational injury implies that the small business employer sees as preventable - and related to exposure - one lost-time injury every second year among his ten workers (in a high risk environment) without considering the reason why the other nine had no injuries during the period. This represents a psychological, an analytical and a micro-economic dilemma.

It is not immediately obvious how, or indeed if, employers differ from the average citizen in their attribution of occupational injury causes and thus would be susceptible to incentives for prevention. Nor is it clear how incentives for prevention would influence investments in safety within the industrial sectors where high physical risks are viewed as normal and technologically determined, or part of the specific sub-culture of hazardous occupational exposure (e.g. forestry, agriculture, fishing, construction, etc).

It is not even clear if size of operation in itself is related to hazard management and the incidence of occupational trauma. A study of explicit and implicit "risk pooling" of small employers in the US workers' compensation insurance market, based on the 1987 National Medical Expenditure Survey, found little support for the hypothesis that employees and dependents of small firms are likely to be averse to health risks should they obtain coverage comparable to that held by large-firm employees (Monheit & Vistnes, 1994).

In 1994 Worksafe Australia funded a study into the relation between direct and indirect cost associated with occupational injury in Australia (Larsson & Betts, 1996). The results of this study indicated that few employers operating a small business meet their obligations under the Accident

Compensation Act as regards reporting and case management. None of the businesses who allowed the researcher to interview the injured employee had reported the injury, and the costs of medical treatments were met by public medical insurance (Medicare).

This implies that there is substantial underreporting of occupational trauma, particularly among small business, in Victoria and, presumably, in Australia. It is also evident that part of the costs associated with the less severe occupational trauma have been shifted from employers and workers' compensation schemes onto the public purse.

A recent study of 100 small business operators in Victoria, Australia showed that small business managers:

- know quite well the types of risks they, their family and staff are exposed to,
- have an intuitive and realistic picture of the relative risk level of their trade in relation to other industries,
- decide on safety investment following the perceived risks, and the severity of consequence (ie. risks of severe or fatal trauma),
- if they belong to an occupational group in a high-risk area, tend to underestimate or be prepared to bear the high inherent health risks of their trade (eg. concreters, farmers). (Larsson, 1998)

The practical, hands-on character of small business ownership, together with the strong involvement of close and extended family will define the perceptions of hazard and risk exposure. For operators of risk-exposed family businesses, safety will be a matter of responsibility to close and extended family.

Lack of knowledge in relation to OH&S laws, regulations and codes is only a problem if such knowledge is a requirement for finding solutions to the risk problems facing the small business operator. It would seem that the present general character of the "duty of care" legal framework is of little practical and applied problem-solving use to the small business operator.

To be relevant and useful to small business, the emphasis of the OH&S information should shift from the legalistic formal administration of safety to the practical solution of typical and high-priority technical risk problems in specific industries. The information included in the claims data of the workers' compensation insurance could potentially be applied for these purposes.

In view of the social characteristics of small business, and their pervasive inclusion of work in family and general life, there would probably be great improvements in successful communication about OH&S if there was a shift in focus from the traditional industrial relations and workers' compensation arguments to a genuine 24-hour perspective of public and family health (Larsson, 1998).

3 Safety – A Market Commodity?

While the proponents of small government and economic rationalism are in power and increasing proportions of social decision-making, together with basic infrastructure in telecommunications, energy and water, are being handed over to different markets, the areas of market failure and social equity will continue to suffer. Weak players in the market - eg. unskilled workers, single mothers, unemployed, the poor, the old and the sick - will be increasingly disadvantaged.

The market demand for safety is patchy and conditional. In some areas this demand has been successfully developed, while in other areas it is only rudimentary. An example of successful, effective and marketable development of safety is represented by the passive and active safety features of the medium-size passenger car.

Although the number of person/kilometers has soared, the chance of surviving a head-on, off-set or side-impact collission as a driver or passenger of a medium-size car has improved manifold (Newstead et al, 2000). For the car market segments in North America and Europe representing family sedans and wagons, active and passive passenger and child safety is a major commercial argument. The safety argument was non-existent in the same market segment in 1970.

But while the car safety development represents a successful market solution, empowering the driver/consumer to gain improved levels of survival and safety, the market power of the risk-exposed, self-employed subcontractor/worker, eg. the electrician, the high-pressure cleaner, the painter, the maintenance repair worker, or the forklift operator, has become more restricted, more exposed to competitive forces, and this has, in some instances, resulted in higher risk levels.

The present model for health and safety legislation and control in the systems dominated by British legal traditions was introduced in a period of strong trade unions, a highly regulated labour market, low unemployment levels, a dominance of standard employment with large employers and well-resourced regulatory agencies. The optimism of "duty of care" regulation can be explained by that structural and political environment.

With de-regulation and reduced government monitoring and control of business activities and the labour market, the informal, grey and black-market part of the economy has consolidated and increased in many areas. The growth of self-employment and small business operations is pronounced in the traditional cash economy industries like retail and trades. In some industrial areas like among owner-drivers in the transport industry, among outworkers in the clothing industry and among subcontractors in the construction industry further structural arrangements compound the growth of black-market work (Clayton, 2000).

Individuals in most modern economies are being asked to assume a greater economic responsibility for their own and their families sickness benefits, to contribute to their social and medical insurances and superannuation schemes (if they live in a jurisdiction which have had such public solutions), and to pay more for their children's education.

If this individual also is a risk-exposed worker his/her typical characteristics in the new economic order are:

- self-employed or in a small (family) business
- low on capital and resources
- merged private/family and business economy
- often relying on some proportion of untaxed turnover
- exposed to competitive tendering
- virtually uncontrolled by safety regulation.

The underlying assumption that risk-exposure is a responsibility for the risk-exposed fits well with this economic and political model. This perspective implies that certain groups in the labour market - performing certain types of work - will be left to control their risks and injury consequences on their own in the competitive market.

Alternatively, one could argue that if the individual is to retain full health through his/her normal span of life then the management of safety at work must be a shared social responsibility. Such a responsibility cannot be devolved to industrial companies or to individual risk-exposed workers. And if this is the case, the social control of occupational safety should be well defined, functionally supported by valid preventative legislation, and underpinned by a well-resourced and competent industrial safety inspectorate.

4 WHAT IS PREVENTION AND HOW IS IT DONE?

The public fund workers' compensation system represents the most valid and reliable way of identifying occupational risks and severe occupational trauma. The claims information of the comprehensive workers' compensation insurer can easily be utilised to identify occupations, practices and exposures which represent unacceptable levels of severe trauma (Larsson, 1990a).

The comprehensive claims information can form the basis for decisions on which industrial exposure warrants control and monitoring, and which equipment, tasks and processes standards and regulations should be explicit and specific (Larsson et al, 1997a; Rechnitzer et al, 1997; Larsson et al, 1997b).

Strategic and long-term prevention activities should be based on the priorities measured by the public fund workers' compensation system, and prevention be directed towards problems representing excessive human suffering, medical impairment or other such severe consequences which are seen as unacceptable by a large majority in society (Larsson & Field, 2000a; Larsson & Field, 2000b; Field et al, 2000).

Systematic injury prevention activities have been successfully performed by a number of organisations in the area of traffic insurance and road safety, eg. the Insurance Institute for Highway Safety in the USA, Folksam Insurance in Sweden, and the Transport Accident Commission (TAC) in Victoria, Australia. The apparent success of Victorian road trauma prevention is an instructive example of how enforcement of easily interpretable rules, pertaining to operator, equipment and system - licencing of drivers, booze buses, car design rules, speed and red light cameras, etc - can be combined with prevention activities directed at specific problems identified by the claims information, eg. black-spots in the road system, young drivers, fatigue, etc. (Corben et al, 1990; Macdonald et al, 1992; Rogerson et al, 1994; Cameron et al, 1993; Newstead et al, 2000).

No-one has suggested that the road traffic regulations should be abolished in favour of a general "duty of care"; the functional enforcement of traffic safety regulation crucial to reduce the risk of fatality and severe trauma on the road is, in fact, supported by most Victorians.
One core feature of applying the information of the public fund workers' compensation system in the area of prevention is that comparisons between perceived local risks and overview information of similar exposures become possible. This means that the data-bank of claims can be restructured into an expert system of information exchange; claims information can be utilized in a system for management decision support on occupational risk (Bråfelt & Larsson, 1993).
For prevention to take place, however, enforcement must be specifically defined and real. But for prevention to be successful it should be kept apart from enforcement.

REFERENCES

ABS, The Australian Bureau of Statistics (1996): *Small Business in Australia.* 1321.0. Canberra.

Bråfelt, O & Larsson, TJ (1993): Sanning och Konsekvens - lokalt skadeförebyggande med hjälp av Trygghetsförsäkringens skadedata. (Truth and Dare - local injury prevention with the help of injury data from the No-Fault Liability Insurance). IPSO Factum 39, Stockholm.

Cameron, MH, Haworth, N, Oxley, J, Newstead, S & Le, T (1993): *Evaluation of Transport Accident Commission Road Safety Television Advertising*. Monash University Accident Research Centre, Report No. 52, Melbourne.

Clayton, A (2000): Occupational Health and Safety Regulation: Challenges for the New Millenium - a Discussion Paper. Draft Document, Monash University Accident Research Centre, Melbourne.

Corben, BF, Ambrose, C & Foong, CW (1990): *Evaluation of Accident Blackspot Treatment*. Monash University Accident Research Centre, Report No. 11, Melbourne.

Cromie, S, Stephenson, B, Monteith, D (1995): The Management of family Firms: An Empirical Investigation. *International Small Business Journal*, **13** (4): 11-34.

Dell, G (2001): *Safety Management: A Hazard Control Function*. Monash University Accident Research Centre, Melbourne.

Dyer, WG (1986): *Cultural Change in Family Firms*. San Fransisco, Jossey Bass.

Field, B, Larsson, TJ, Kenningham, L & Lee, S (2000): Strategic Occupational Injury Prevention: Falls from Heights - A Report on Claims, Fatalities and Injury Severity Outcomes. *Policy Research Report* No **8**, Part 3. Victorian Workcover Authority, Melbourne.

Goffee, R, Scase, R (1985): Proprietorial Control in Family Firms. *Journal of Management Studies*, **22**: 53-68.

Gun, R (1992): Regulation or Self-Regulation: Is Robens-Style Legislation a Formula for Success? *Journal of Occupational Health and Safety - ANZ*, **8** (5): 383-8.

HSO, Health and Safety Organisation (1995): *Business Plan 1995/96*. Government of Victoria, Melbourne.

Holmes, N (1994): *OHS Promotion and Risk Communication in Small Business.* Belts to Bytes Factories Act Centenary Conference, Workshop Papers, pp 29-42, Workcover Corporation, South Australia.

Johnstone, R (1994): *The Court and the Factory: The Legal Construction of Occupational Health and Safety in Victoria.* Thesis, University of Melbourne.

Larsson, TJ (1990a): *Accident information and priorities for injury prevention* (Doctoral Dissertation, Royal Institute of Technology, Stockholm). IPSO Factum 21, Stockholm.

Larsson, TJ (1990b): Severe Hand Injuries among Swedish Farmers. *Journal of Occupational Accidents*, **12**: 295-306.

Larsson, TJ (1997): Inspection and Prevention: Present Concepts of Occupational Injury Prevention in Scandinavia and Australia. *Safety Science Monitor*, **1**(1), www.ipso.asn.au.

Larsson, TJ (1998): Decision Making in Relation to Occupational Health & Safety among Small Business. A Survey of 100 Small Business Owners/Managers in Victoria. *Policy Research Report* No. **7**. Victorian Workcover Authority, Melbourne.

Larsson, TJ, Rechnitzer, G, Ozanne-Smith, J, Newstead, S & Gantzer, S (1997a): Strategic Occupational Injury Prevention: Occupational Black Spots - Victoria (Analysis and Priorities). *Policy Report* No **6**, Part 1. Victorian Workcover Authority, Melbourne.

Larsson, TJ, Rechnitzer, G & Lee, S (1997b): Strategic Occupational Injury Prevention: “Operation Safety” - Preventing Occupational Injuries Through Regional Intervention. *Policy Report* No **6**, Part 3. Victorian Workcover Authority, Melbourne.

Larsson, TJ & Betts, N (1996): The Variation of Occupational Injury Cost in Australia; Estimates Based on a Small Empirical Study. *Safety Science*, **24** (2): 143-55.

Larsson, TJ & Field, B (2000a): Strategic Occupational Injury Prevention: The Distribution of Occupational Injury Risks in the State of Victoria. *Policy Research Report* No **8**, Part 1. Victorian Workcover Authority, Melbourne.

Larsson, TJ & Field, B (2000b): Strategic Occupational Injury Prevention: The Distribution of Occupational Injury Risks in the Victorian Construction Industry. *Policy Research Report* No **8**, Part 2. Victorian Workcover Authority, Melbourne.

Leach, P (1991): *The Family Business.* London, Kogan Page.

Macdonald, WA, Bowland, L & Triggs, TJ (1992): *Evaluation of Driver Testing and Licensing in Victoria.* Monash University Accident Research Centre, Report No. 24, Melbourne.

Mayhew, C (1997): *Barriers to Implementation of Known Occupational Health and Safety Solutions in Small Business.* Worksafe Australia and the Queensland Government Division of Workplace Health and Safety.

Mayhew, C, Young, C, Ferris, R, Harnett, C (1997): *An Evaluation of the Impact of Targeted Interventions on the OHS Behavior of Small Business Building Industry Owners/Managers/Contractors.* Department of Training and Industrial Relations Queensland Government and the National Occupational Health and Safety Commission.

Monheit, AC, Vistnes, JP (1994): Implicit Pooling of Workers from Large and Small Firms. *Health Affairs*, **13** (1): 301-14.

Morrisey, MA, Jensen, GA, Morlock, RJ (1994): Small Employers and the Health Insurance Market. *Health Affairs*, **13** (5): 149-61.

Mäkinen, H (1982): *Evaluation of the Effects of Press Regulations in Finland.* Nordiska Olycksfallsforskningsseminariet 18-20 augusti 1981. OARU, Royal Institute of Technology, Stockholm.

Newstead, S, Cameron, MH & Le, CM (2000): *Vehicle Crashworthiness and Agressivity Ratings and Crashworthiness by Year of Vehicle Manufacture: Victoria and NSW Crashes During 1987-98, Queensland Crashes During 1991-98.* Monash University Accident Research Centre, Report No. 171, Melbourne.

Oxenburgh, M (1991): *Increasing Productivity and Profit through Health & Safety.* CCH International, Sydney.

Pettersson, B (1982): *Olycksfallsutvecklingen i skogsbruket under 1970-talet.* Nordiska Olycksfallsforskningsseminariet 18-20 augusti 1981. OARU, Royal Institute of Technology, Stockholm.

Quinn, JJ (1997): Personal Ethics and Business Ethics: The Ethical Attitudes of Owner/Managers of Small Business. *Journal of Business Ethics*, **16**: 119-127.

Rechnitzer, G, Larsson, TJ & Mantle, B (1997): Strategic Occupational Injury Prevention: Occupational Black Spots - Ballarat Region. *Policy Report* No **6**, Part 2. Victorian Workcover Authority, Melbourne.

Lord Robens (1972): *Safety and Health at Work, Report of the Committee, 1970-72.* London: HMSO, CMND 5034 (Roben Report).

Rogerson, PA, Newstead, S & Cameron, MH (1994): *Evaluation of the Speed Camera Program in Victoria 1990-91.* Moansh University Accident Research Centre, Report No. 54, Melbourne.

Smith, KG, Gannon, MJ, Grimm, C, Mitchell, TR (1988): Decision Making Behavior in Smaller and Larger Professionally Managed Firms. Journal of Business Venturing, **3** (3): 223-32.

Springfeldt, B (1996): Rollover of tractors - international experiences. *Safety Science.* Vol **24** (2):95-110.

Stanley, P, Shikdar, A, Cross, J, Gardner, D, Carlopio, J (1997): *Mechanical Equipment Injuries in Small Business.* Department of Safety Science, University of NSW.

VWA Advisory Committee (1994): *Incentives for Accident Prevention Among Small and Medium Sized Employers.* Victorian Workcover Authority.

Whisler, TL (1988): The Role of the Board in the Threshold Firm. *Family Business Review*, **61**: 143-54.

Our second contribution from Australia is by Hopkins. It moves to the other end of the size spectrum and considers the regulation of large companies operating in highly hazardous environments.

The problems of regulating highly hazardous operations which may impact on the wider public as well as employees has received considerable attention in recent decades, symbolised by the adoption of the Seveso directive in Europe in 1982, requiring operators to make a "safety case" to the regulator. Hopkins argues that a safety case runs the risk of being a static, once-and-for-all activity which is implicitly assumed to guaranteed safety thereafter. He describes an alternative model for the regulation of major hazards, drawn from the coal industry in the state of Queensland, which stresses ongoing mindfulness of the possibility of disaster and the need for companies to define trigger events which, if they occur, will lead to rapid and effective intervention before problems get out of hand. This idea is consistent with contemporary thinking about the mindfulness of high reliability organisations.

3

Two Models of Major Hazard Regulation: Recent Australian Experience

Andrew Hopkins

Recent disasters in Australia in different industries have prompted different regulatory responses. In the state of Victoria, a gas plant accident which disrupted supplies to the state for two weeks (Hopkins, 2000) has led to the introduction of a safety case regime, while in the state of Queensland, a coal mine explosion which killed eleven men (Hopkins, 1999) led to a requirement that mines develop principle hazard management plans. This paper looks at the characteristics of the two different styles of regulation, explores reasons for these different responses, and draws some tentative policy conclusions.

The Victorian Safety Case Regime

Safety case regulation modelled on the Seveso directives has been developing in Australia in recent years. From about 1995 Australian offshore petroleum production has been subject to a safety case requirement, while on shore, a major federal government report in 1996 recommended that the Australian states should introduce safety case requirements for major hazard facilities (NOHSC, 1996). Little progress had been made in implementing this recommendation by the time of the accident at the Esso Longford gas plant in Victoria in 1998. A Royal Commission of inquiry into this accident recommended the implementation of a safety case regime for major hazard facilities in Victoria (Dawson, 1999) and the state government has now enacted the necessary regulations (Victoria, 2000).

The Victorian regulations draw explicitly on the Seveso II directive (see Porter and Wettig, 1999). The centrepiece is the requirement that major hazard facilities present a safety case or report to the appropriate government authority.

Under the regulations the operator must carry out a formal safety assessment. This involves, first, systematically identifying all major hazards and major accidents which could occur, second, assessing the possible consequences and third, specifying control measures. Details of the formal safety assessment, including the hazard identification methodology, risk assessment criteria, and justification for control measures used, must be included in the safety case. The regulations do not specify the methodologies to be used in identifying and controlling hazards.

The operator must also establish a safety management system and a summary of the content of this system must be provided in the safety case. The regulations detail a number of things which must be included in the safety management system. It is the function of the safety management system to ensure the controls envisaged by the formal safety assessment remain effectively in place.

Once the authority is satisfied that the case complies with the regulatory requirements it must issue a license to operate. Major hazard facilities may not operate without such a license. Licenses are valid for up to five years after which the safety case must be revised, and revisions can also be triggered by changed circumstances.

There is a strong emphasis in this approach on 'getting in right in the first place'. The Victorian Government has estimated that most of the cost associated with meeting the safety case requirements occurs prior to licensing (VWA, 1999b). There is an implicit presumption that once the license has been obtained, the ongoing process of safety management becomes more straightforward. It is certainly not the intention of the regulations that operators pay less attention to safety after they have been licensed, but there is a danger that that is how the regulations will be interpreted by those who manage these facilities. This danger can be avoided if regulators carry out intensive and ongoing audits of safety management systems to ensure that they are functioning in practice as they are intended to on paper. I have stressed this point, however, in order to contrast the safety case strategy with the alternative approach to be discussed below.

THE QUEENSLAND COAL INDUSTRY REGIME

Coal mining in Australia has traditionally been subject to industry-specific safety legislation. In the state of Queensland, following the Moura mine disaster, the legislation was entirely rewritten and a number of approved standards were promulgated. The legislation defines a principle hazard at a coalmine as one with the potential to cause multiple fatalities. It requires each mine to develop a safety management system which includes management plans for each principle hazard. The generic approved standard (QDME, 1998) specifies that, at a minimum, the following six principle hazard management plans must be developed:

- Ventilation management
- Gas management
- Emergency evacuation
- Methane drainage
- Spontaneous combustion
- Strata management (roof control)

Examination of this list reveals that it is not simply a list of six hazards for which plans must be developed. It refers to both specific hazards (e.g. spontaneous combustion) and general control systems (e.g. ventilation management), which may be applicable to more than one hazard. The principle hazards are to some extent implicit in this list. They include methane gas explosions (which may or may not trigger coal dust explosions), carbon monoxide poisoning, roof or wall collapse, and the long latency period dust disease, pneumoconiosis, which has probably killed more miners in the long run than anything else.

One difference between this and the safety case approach is already apparent. Whereas the safety case approach puts great emphasis on facility operators carrying out their own hazard identification, and doing it as systematically and transparently as possible, under the Queensland coal industry model, the authorities have themselves implicitly identified the most important hazards which need to be controlled.

I focus in what follows on the approved standard for a particular hazard, spontaneous combustion, since this phenomenon has been responsible for three major explosions in underground coal mines in Queensland in a space of 20 years. A total of 41 men were killed in these explosions.

Once exposed to air, coal has a tendency to oxidise and to heat up slowly, in a process known as spontaneous combustion. Unless it is well ventilated the temperature can rise to the ignition point of methane. Should a combustible quantity of methane be present, it will explode. When spontaneous combustion is occurring it gives off carbon monoxide (CO), and underground coal mines are equipped with gas monitoring systems to measure, among other things, the quantity of CO being produced. The greater the quantity, the greater the cause for concern.

The approved standard states that plans must specify "trigger" events which will require further action. It does not state explicitly that carbon monoxide must be used as the basis for setting triggers, but this in fact is industry practice. For example, one mine, to be discussed below, sets three trigger levels for CO concentrations in areas where mining has been completed and which are sealed off (except for gas monitoring points). They are as follows: less than 50 ppm (parts per million) is normal; 50-120 ppm is a level 1 alarm; between 120-500 ppm is a level 2 alarm and above 500 ppm is a level 3 alarm.

The approved standard requires, further, that plans specify the action to be taken if trigger levels are reached. The main action required is the establishment of a control group of predetermined composition to manage the event. The group must contain people of sufficient authority to implement decisions, it must maintain a log of its activities and it must not disband until the situation has been resolved. At the mine in question the composition of the group depends on the alarm level. At level 1 the control group may consist of the mine manager alone. At level 3, on the other hand, management has decided that the control group must contain people from outside the mine, including someone from corporate headquarters, a government inspector and a full-time union safety officer.

A Spontaneous Combustion Management Plan in Action

A major spontaneous combustion event occurred at this mine soon after it had adopted its plan, and it is worth recounting the details (Stephan, personal communication). CO monitoring picked up indications that spontaneous combustion might be occurring in a mined out and closed off area of the mine. The readings were high enough to trigger a level 3 alarm which, among other things, required the withdrawal of men from the mine. The control group was constituted, as described above. After some hours management came to the conclusion that the readings did not indicate a current problem and that the gas sampling process in the closed off area was picking up remnants from an incident which had occurred five months earlier. The mine manager was concerned about lost production and wanted the men to return to work underground as soon as possible. He put considerable pressure on the control group to endorse his view, but the external members held out against this pressure. An external expert was then called in to review the gas readings and he concluded that the situation was urgent, that a fire was burning somewhere underground, and that unless it was extinguished immediately an explosion was highly likely. At this point management acted decisively, hired an inert gas generator and pumped the affected part of the mine full of inert gas. CO readings indicated that the fire was extinguished within a matter of hours.

The outcome was a vindication of the mine's spontaneous combustion management plan. Moreover, in a subsequent analysis (Stephan, 1998) the government inspectorate stressed that the experience had highlighted the importance of including outside personnel in the incident control group There is enormous pressure on control group members to interpret ambiguous indicators in such a way as to allow production to continue. The co-option of group members to management's viewpoint was only prevented by the presence of outside participants, in particular, the union safety official. This was a situation where the well know phenomenon of groupthink (Janis, 1972) might have been expected to operate. One of the techniques by which groupthink can be overcome is if someone is given the role of devil's advocate with the job of raising doubts about any decision which the group appears likely to make. In the above case the union official, quite naturally, acted as a devil's advocate and this was crucial in ensuring that the control group functioned optimally.

Identifying Warnings

The essential feature of the spontaneous combustion management plan is the idea that certain events or conditions constitute a warning that something might be amiss and are to be treated as triggers to action. The action may simply be to investigate the situation, but in more serious cases it may involve the cessation of production until the matter is resolved. It is easy to seen how the notion of trigger events applies to the control of spontaneous combustion. But the strategy is more general. The generic standard under Queensland coal mine safety legislation requires *all* principle hazard management plans to identify triggers which require specified action (QDME, 1998:12). In particular, it requires that triggers for the withdrawal of persons must be established for each identified principle hazard (QDME, 1998:39). The thinking behind this approach is that major accidents in coal mines are always preceded by warnings or indicators that something is amiss. The aim of the principle hazard management plan is to systematically identify these warnings and ensure that appropriate action is taken, before any major incident occurs.

EXPLAINING THE DIFFERENT REGULATORY RESPONSES

Why is it that these two disasters generated such different regulatory responses? Is it because the industries are so different?

One way in which gas refining and coal mining differ is in the relative difficulty of identifying major hazards. In underground coal mining, the major hazards are well known; the problem is to control them. On the other hand, in gas processing plants the biggest problem is to identify the hazards. As Esso's parent company, Exxon, has commented, "once the hazards have been identified, a major stumbling block to loss or accident prevention has been overcome" (Dawson, 13.50) A safety case regime emphasises the importance of identifying hazards and is thus, from this point of view, an appropriate style of major hazard regulation for gas plants. This emphasis would be somewhat misplaced in the underground coal industry, where the hazards are relatively well known. This is perhaps one reason why the Queensland authorities did not adopt the safety case approach in designing the new major hazard regulatory regime for the underground coal mining industry.

Second, in the case of chemical plants, it is often possible to eliminate hazards at source by careful design of the system. The safety case approach stresses this idea of 'getting it right in the first place' and is therefore particularly appropriate. In contrast, the hazards of coal mining are not man-made and are not as easily eliminated. The emphasis must therefore be on the ongoing management of these hazards, for which the trigger event model is particularly appropriate.

There is a third and related reason why the trigger event model is particularly suited to the coal industry.

A mine's operating environment changes constantly as mining progresses. Roof strengths may vary, requiring changes in roof support strategy; the gas produced by the coal may change because of the presence of faults and other geological discontinuities and this may require changes in ventilation flows, and so on. The trigger event mechanism in the principle hazard management plans allows the levels of control to be varied as circumstances change. It would, for instance, be inefficient to require roof support of the highest level to be installed when the roof in most cases does not require it. (Stephan, personal communication)

A continuously changing physical environment creates a need for constant alertness to danger and the trigger event strategy is one way of promoting this alertness.

A fourth, potentially relevant way of looking at the difference between the two industries concerns degree of coupling (Perrow, 1999). Gas processing plants are tightly coupled, in the sense that when something goes wrong in one part of the system it may propagate rapidly to other parts of the system with little opportunity for human intervention before disaster occurs. On the other hand, coal mining, at least according to Perrow (1999:245), is more loosely coupled. Accident sequences develop over a longer period of time and therefore offer more opportunity to intervene. On this basis it might be thought that the regime adopted for coal mining in Queensland is particularly suited to that environment and might not be as applicable to more tightly couple environments, such as gas plants. The safety case concept was developed with tightly coupled systems in mind and was the natural regulatory response to the Longford gas plant accident.

However, the coupling argument is easily overstated. Even in tightly coupled environments, accident sequences often take many hours to develop. The Longford accident was initiated by events occurring nearly a day before the final explosion, and perhaps more importantly, the antecedent causes of the event went back years. There were numerous points at which human intervention might have averted the accident, until almost the last moment. The potential for responding to whatever warning signs there are remains considerable, even in tightly coupled systems.

APPLYING THE TRIGGER EVENT MODEL MORE WIDELY

The preceding discussion suggests the safety case model is not as appropriate for the coal industry as the trigger event model. The intriguing question remains, however, of whether the trigger event model might be applicable in industries currently governed by safety case regimes. Barry Turner's theory of accident causation, propounded in *Man-Made Accidents*, suggests that the answer is yes. On the basis of a wide-ranging review of major accident reports, he concludes that disasters always involve an information or communication failure of some kind. There is always information available somewhere in the organisation that constitutes a warning which, if heeded, would have averted the accident. For Turner, the central question to ask, then, is: what stops people from acquiring and using the appropriate advance warning information so that large-scale accidents and disasters are prevented?" (1978:195). He provides a variety of answers:

1. Information is noted but not fully appreciated. This may occur for a number of reasons, including a false sense of security which leads individuals to discount danger signs, pressure of work which diverts attention from warning signs, or difficulty in sifting the information from a mass of other irrelevant facts.
2. Prior information may not be correctly assembled. This may be because it is buried in other material, distributed among several organisations or distributed among different parties within the same organisation.

3. Bad news is usually unwelcome. As a result, people in possession of relevant warning information may be disinclined to pass it up the line to senior managers.

Based on this analysis it would seem that the starting point for effectively managing any major hazard must an incident reporting system. But if this is to have any chance of gathering relevant warning signs, management must put considerable thought into specifying what sorts of things should be reported: what are the warning signs that something might be about to go disastrously wrong? Events which management might decide to treat as warnings include: certain kinds of leaks; certain kinds of alarms; particular temperature, pressure or other readings; certain maintenance problems; machinery in a dangerous condition and so forth. Management must ensure that such information is indeed reported or collected. Once identified, these reported events must be treated as triggers to action and management must specify what kinds of action is required and who is responsible for taking the action. For instance: should a further investigation be undertaken? should production be stopped? Such a system structures decision making in a way which forces organisations to take note of and respond to warnings of danger.

The strategy just described is essentially the trigger event model adopted by the Queensland coal industry. In principle there is no reason why it could not be applied in other industries.

An objection which is sometimes made to this approach is that it is only obvious with hindsight that events constituted warnings of what was about to happen. How can warnings be distinguished from all the other signals in the environment which amount to no more than noise? The answer is that the trigger event model does not purport to identify beforehand which are the real warnings and which the false alarms. Instead it identifies certain categories of event which at least call for further investigation. That investigation may reveal that the matter is of no consequence. But the beauty of the system is that it ensures that such a judgement is made consciously and conscientiously and does not simply occur by default.

APPLYING THE TWO MODELS TO THE LONGFORD ACCIDENT

I want now to illustrate the differences between these two approaches by showing how each might have worked to prevent the Longford accident. That accident occurred because certain warm oil pumps failed, causing a large metal heat exchanger to be chilled to well below its design temperature. As a result, ice formed on exterior pipe-work which was normally too hot to touch and leaks developed because of metal contraction. The metal became brittle with the cold. The operators did not understand the dangers of embrittlement. After some time they succeeded in restarting the warm oil pumps. The metal vessel shattered, releasing a large volume of volatile hydrocarbons which subsequently ignited. This hazard of cold metal embrittlement had not previously been identified by the company.

Consider how a safety case regime would have prevented this accident. The standard hazard identification procedure in the industry is a HAZOP (hazard and operability study). Two other gas plants at the Longford site had been HAZOPed but a planned HAZOP of the plant where the accident happened had been deferred indefinitely. A HAZOP would almost certainly have identified the possibility that a failure of the warm oil pumps might result in the embrittlement of certain pressure vessels and controls would have been put in place to ensure that this did not happen. A safety case regime would have required the company to demonstrate that it had identified and controlled all hazards, and it is most unlikely that the regulatory authority would have accepted a safety case in which a HAZOP had not been carried out at one of the gas plants. In short, had a safety case regime been in place, the hazard would have been identified, assessed and controlled from the outset.

Consider, now, how a strategy of identifying and responding to warning signs might have averted the accident. One month prior to the accident the main warm oil pump and its backup had to be shut down simultaneously for maintenance. This had never happened before. Just as described above, the metal heat exchanger became abnormally cold, ice formed on exterior pipe-work which was normally too hot to touch, and the heat exchanger sprang a substantial leak. More by good luck than good management the metal heat exchanger did not get to the point of embrittlement and when operators restarted the warm oil flow, the system returned to normal. The event produced clear warning signs that something was amiss. But their significance was not understood and they were ignored. In particular, the matter was not recorded in the company's incident reporting system, because the reporting system was used primarily to report personal injuries. Had the event been entered into the reporting system it would have come to the attention of more senior company managers and a root cause analysis would have be carried out, in accordance with company policy. This would have identified the danger, and measures to control it would then have been instituted.

It is clear, therefore, that both styles of major hazard regulation would have averted the accident, had they been in place. A safety case regime would have mandated a systematic hazard identification procedure which would have identified and controlled the hazard of cold temperature embrittlement. An incident report system which required the reporting and investigation of abnormal temperature events and leaks would also have resulted the discovery and control of the danger of embrittlement.

It might be objected that in all the years of operation at Longford there had only been one warning event prior to the accident. Had this event not occurred the regulatory strategy which relied on responding to warnings would have failed. However the parent company, Exxon, had considerable previous experience of brittle fracture. Had Esso Australia been required to respond to warnings elsewhere in the company, it would have been alerted to the danger and responded accordingly. There may a jurisdictional problem here, but if this can be overcome, a regulatory regime based on responding to warnings would appear to be a viable alternative to a safety case regime.

If organisations are willing to look beyond their own immediate confines, the possibility of identifying and responding to relevant warnings is enormously expanded. To give just one example, the near disaster at the Three Mile Island nuclear power station was preceded, 18 months earlier, by an almost identical accident sequence at another nuclear power plant in the US, at Davis-Besse (Gorinson, 1979; Hopkins, forthcoming). Information about this event was available to relevant decision makers at Three Mile Island but it was not attended to. Had it been, plant operators would have behaved differently and the accident would have been averted.

CONCLUSION

Recent research suggests that organisations that function with high reliability are characterised by collective mindfulness (Weick et al, 1999). They put in place systems which promote a general wariness and a suspicion of apparently incident-free operation. They worry about the possibility of failure and are acutely attentive to indications of danger. It is clear that the principle hazard management plans of the Queensland coal industry, with their focus on trigger events, promote this ongoing collective mindfulness. Their assumption is that safety is never guaranteed once and for all but depends on continual organisational attentiveness to the possibility of disaster. In contrast, the safety case approach to the management of major hazards, with its emphasis on getting it right in the first place, is not as inherently conducive to continued mindfulness of the possibilities of disaster.

This does not mean that the trigger event model should replace the safety case, for the two models are not incompatible. It would be quite possible to modify the safety case approach by incorporating a trigger event requirement. Safety case regulations could require that facility operators identify the kind of information which will count as a warning of danger and hence as a trigger to action. They would be required to show that they had systems in place for ensuring that this information was gathered effectively and responded to conscientiously.

It is interesting to note that Victorian safety case regulations already contain some elements of this solution. One of schedules attached to the regulations requires that safety cases identify "critical operating parameters" (e.g. temperature or pressure limits which should not be breached), while another requires that safety cases identify performance measures which include failures to keep a process within the critical operating parameters. The regulations do not, however, treat these events as triggers which require action. It would be easy enough to include such a requirement in the regulations. Giving these elements sufficient prominence would then effectively introduce the trigger event strategy into the regulations. It is clear that the trigger event approach is a potentially important supplement to the safety case strategy, promoting the collective mindfulness which is necessary for reliable and safe operation.

REFERENCES

Dawson, D., & Brooks, B. (1999). *Report of the Longford Royal Commission: The Esso Longford Gas Plant Accident*. Melbourne: Victorian Govt Printer.

Gorinson, S. et al (1979). *Report of the Office of Chief Counsel on the Role of the Managing Utility and its Suppliers*. Washington: US Govt

Hopkins, A. (1999). *Managing Major Hazards: The Lessons of the Moura Mine Disaster.* Sydney: Allen & Unwin.

Hopkins, A, (2000). *Lessons form Longford: The Esso Gas Plant Explosion.* Sydney:CCH

Hopkins (forthcoming) "Was Three Mile Island a Normal Accident?"

Janis, I. (1972). *Victims of Groupthink: A Psychological Study of Foreign-Policy Decisions and Fiascoes.* Boston: Houghton Mifflin

NOHSC (National Occupational Health and Safety Commission). (1996a). *Control of Major Hazard Facilities: National Standard.* Canberra: AGPS.

Perrow, C. (1999). *Normal Accidents (With New Afterword).* Princeton: Princeton U.P.

Porter, S., & Wettig, J. (1999). Policy issues on the control of major accident hazards and the new Seveso II directive. *Journal of Hazardous Materials*, **65**, 1-14.

QDME (Queensland Department of Mines and Energy), (1998) *Approved Standard for Mine Safety Management Plans.* Brisbane: DME

Stephan, S (1998) "Decision making in incident control teams". Paper to the Queensland Mining Industry OH&S Conference.

Turner, B. (1978). *Man-Made Disasters* . London: Wykeham.

VWA (Victorian WorkCover Authority). (1996). *OHS Management System Audit of Esso Australia* .

Victoria (2000). *Occupational Health and Safety (Major Hazard Facilities) Regulations 2000.* Melbourne: Government printer.

VWA (Victorian WorkCover Authority). (1999). *Proposed Occupational Health and Safety (Major Hazard Facilities) Regulations: Regulatory Impact Statement.* Melbourne: VWA.

Weick, K., Sutcliffe, K., & Obstfeld. (1999). Organising for high reliability: processes of collective mindfulness. *Research in Organisational Behaviour*, **21**, 81-123.

The next contribution, by Oh, shows how the risk of complacency in safety case systems, identified in Hopkins' chapter, is being overcome in the European context, by emphasising the importance of ongoing safety management systems

The Seveso II directive is generally regarded as the best model available for the regulation of major hazards. Oh describes the evolution from Seveso I to Seveso II and he speaks with authority, as a Dutch regulator who was centrally involved in the development of Seveso II. The requirement of Seveso I was that facility operators "tell us" (the regulator) that they are safe; Seveso II requires them "to demonstrate to us you are safe". Many of the industries described in this volume are subject to safety case regimes and Oh's paper therefore provides the background which helps us understand this relatively new regulatory style.

Oh describes the process of implementing Seveso II in Holland and it becomes clear that it requires a variety of regulators - labour, environmental, fire safety and others to work together. This is a challenge to regulators which has not hitherto been much commented on. It creates considerable difficulties but also presents opportunities for cross-fertilisation and improvement in regulatory practice.

One of the themes of this book is the changing role of the regulator as we move for prescriptive to goal setting regulation. This is normally assumed to mean that regulators will be less involved in direct supervision. Oh argues that on the contrary the Seveso II directive draws regulators into a role where they are involved in and become to some extent responsible for maintaining safety. This is perhaps one of the ironies of the new arrangements.

4

THE EU SEVESO II DIRECTIVE: AN EXAMPLE OF A REGULATION THAT COULD ACT AS AN INITIATOR TO RAISE THE MAJOR HAZARD SAFETY AWARENESS WITHIN SOCIETY

J.I.H. Oh

INTRODUCTION

In the past decades technology and especially chemical technology has developed rapidly. Synthetic equivalents were developed for products like wool, linen, cotton and rubber, and new products like insecticides, pesticides, medicine and fertilisers were introduced. Moreover, the demand for these products increased. This has had the following effects: continuous growth of production installations, storage and transport of dangerous goods on a large scale, more frequent use of hazardous substances and the development of complex installations with more extreme and critical process conditions. The chemical process industry still has the reputation of being a very safe industry. However, although the probability of an accident is very low, the consequences are severe. This was shown in 1974 Flixborough UK (28 killed), 1975 Beek the Netherlands (14 killed) and last but not least 1976 Seveso Italy. These accidents in combination with the rather arrogant attitude of the industry led to a change of perception. Society perceived the risks of chemical industry as high and it was only a matter of time before politicians decided that within the chemical process industry a more active prevention policy was necessary.

POLICY DEVELOPMENT

Laws and regulation in various countries and the way they are adhered to can be totally different depending on the country, the culture and the regulation style. In the Scandinavian countries safety management systems always have been very important in that technical codes and standards are secondary to safety management regulations. This is very goal orientated regulation. Germany is at the other end of the scale. There the view is that if installations are built to be technically perfect, there is no need for regulation with respect to safety management systems. So their regulation is very prescriptive. The UK and the Netherlands have a mix of the two types of regulations and are in fact very similar to each other.

In the Netherlands, the Safety Act 1934 was amended in 1977 by introducing the obligation for companies that were dealing with large amounts of hazardous substances to submit a safety report in which they were obliged to give descriptions of their activities, the installations, processes, hazards, safety measures, safety policy and safety organisation etc. This was in essence goal directed regulation. It was goal directed rather then prescriptive because it was recognised that modern installations, processes and organisations had become complex to a degree that it was becoming extremely difficult to regulate safety in a deterministic, prescriptive way.

Totally new in this amended legislation was the obligation for companies to perform a series of safety studies. The more potentially hazardous the installation, the more extensive the demanded study, resulting ultimately in forcing the company to do HAZOPS (hazard and operability studies). Certainly in those days, demanding HAZOPS in regulation was a cutting edge. Another new aspect was the so called system approach. This means that during the safety assessment by the inspectorate the total safety situation, both technical and organisational, is assessed in a holistic way. The assessment is tailor-made, rather then generic. The idea behind this regulation was twofold: to make employers aware of the hazards that they are dealing with and to give the government insight into the major hazards of these particular companies.

Within the Netherlands with respect to major hazard safety, there are two approaches. Environmental law embodies the first and is based on the concept of prescriptive regulation in combination with codes and standards and a licensing system. The local authority will issue the license and is completely independent from central government. Law enforcement takes place on the basis of the licence. If a company is inspected and the situation differs from the one in the licence, the company has to make corrections in their facility. Regulation is strongly based on separation and distances based on risk models and is more preventive then repressive.

In 1982 the Safety Act 1934 was replaced by the Working Conditions Act. The Working Conditions Act regulates the relation between employer and employee. The regulator is the third party who enforces the law The second approach referred to above occurs in the Working

Conditions Act. This is a goal setting regulation. In this law the boundaries are set within which the companies have to operate. There is no licensing and companies are free to do whatever they want as long as they stay within the context of the law. Enforcement takes place by inspections and if the inspector finds that a situation is not within the boundary of the law, the company has to change it. To be able to determine how legal requirements should be met, regulator and industry used to work closely together. In this consensus model, industry would draw up their own codes and standards under close supervision of the regulator. Consensus means that on the appropriate levels, from political via policy to technical level, committees or working parties have to be drawn up to ensure commitment all the way. One of the weaknesses of such a system is the lack of co-ordination. Every party involved: regulator, employers and workers have to see to it that the efforts are properly co-ordinated. In practice, this has been difficult.

At the time that this Act was set up, a whole series of committees were installed to ensure consensus between regulator and industry. These included: the existing Committee for Prevention of Disasters by Dangerous substances, an inter departmental committee that draws up directives with respect to the storage and handling of dangerous substances; a tri-partite "Informal Committee" which in fact was very formal and which dealt with policy and regulatory issues; a committee called "RiVePro" which consisted of regulator and technical professionals from the industry which drew up state of the art guidance on all sorts of subjects related to safety in the process industry; and a committee that discussed all the problems with the implementation of the aforementioned regulation. This series of committees saw to it that communication between regulator and industry was optimal. It is interesting to note that all the parties involved recognised the importance of this communication. The result of this approach was a very modern act which was ahead of its time. It was ahead of its time because for the first time a holistic system approach was used. The safety system was judged as a whole, rather then, as was more common in those days, on all the individual technical bits and pieces. This was an integrated approach, where safety management and technical standards together were judged on their consistency and robustness. For the first time in the history of the labour inspectorate, all the technical specialist inspectors were being trained in safety auditing and all sorts of other social skills.
In my view this Act was the locomotive which helped to advance the safety awareness in the process industry in the Netherlands.

INTERNATIONAL DEVELOPMENTS

Parallel to these Dutch developments policy was developed with respect to major hazards both within the European Commission and several member states. This resulted in, amongst others, the EU Seveso directive, the German Störfall-Verordnung, the UK CIMAH regulations and the ILO 174 Major Hazard convention .

During the debate within the EU it was decided that the Seveso directive would be an Environmental (DG XI) directive which would deal with the hazards for man and environment. Furthermore it was decided to establish a committee of technical experts from all member states chaired by the Commission. In this committee (Committee of Competent Authorities, CCA) proposals for the directive were discussed, and in a later phase, so were the implementation issues and monitoring of the state of play within the member states. Gradually this CCA has evolved to become an important forum where, besides the commission and EU member states, participants include the accession and pre-accession countries, the OECD, the Major Accident Hazards Bureau and the EU directorate responsible for research. The Major Accident Hazards Bureau (MAHB) is part of the EU Joint Research Centre and plays an important role with respect to providing scientific support.

THE SEVESO DIRECTIVE

In 1982 after about three years of negotiation the EU adopted the Seveso directive (82/501/EEC). The directive is aimed at a minimum standard of harmonisation of major hazard regulation in member states. It was very much aimed at providing the authorities with information about the safety of a site, based on a safety report that had to be submitted by the company. Central in the regulation was the handling of dangerous substances. The directive included a system of threshold values of dangerous substances. If these values were exceeded a company had to draw up a safety report which basically described all the processes and safety systems connected to the dangerous substances and the measures taken to prevent loss of containment and the system of mitigation. As such this directive was an administrative directive, not setting any safety level but forcing the companies to describe in great detail what they were doing and what measures were being taken to prevent and mitigate loss of containment. Also this directive deals almost exclusively with the dangers for humans and not the dangers for the environment.

The experts in the CCA of the more advanced countries in this area were concerned that the standard of the EU would be set too low. It was therefore a pleasant surprise that this did not happen. On the contrary, standards were set in line with legislation of those more advanced countries. Discussions within the CCA were not only technically but also politically influenced, which resulted in a directive which in certain aspects were difficult to explain .

A first small step was to centralise European major hazard policy with articles in the directive giving the Commission co-ordinating powers with respect to the monitoring and reporting of the implementation in different countries and the reporting of incidents by the member states to the Commission. The directive was amended twice. The first time after the Bhopal accident (1984, 2500 people died) and the second time after the Sandoz accident (1986, 500.000 fish died).

In the Netherlands the directive was implemented by the Ministry of Social Affairs and Employment (the existing safety report regulation), the Ministry of the Environment (new external safety report regulation) and the Ministry of the Interior (new emergency planning regulation). The end result of this implementation was that it exceeded the European directive with respect to demands towards Safety Management Systems (Ministry of Social Affairs) and Quantified Risk Assessment (Ministry of the Environment). Obviously this gave rise to extensive discussions with industry and led to discussions in parliament. As a result, during new policy development the ministries had to involve industry to a greater extent then normal.

It is interesting to note all the differences between European member states. Every country seems to have a regulatory culture of their own and a reputation of their own. In general southern European countries are thought to be less good in implementing European directives then their Northern colleagues. Comparison of implementation track records however shows that southern European countries have the same or sometimes even better percentage of totally implementing European directives. There is a suggestion that law enforcement differs greatly over the European countries. To my knowledge there are no scientific studies that underpin such a statement.

THE SEVESO II DIRECTIVE

In 1950-1960 the chemical industry went through a fast development phase. Technology was new and business was booming. Because of this new high tech technology nobody in the outside world knew exactly what was going on. This was also stimulated because of the rather arrogant and closed behaviour of the chemical industry. They had what can be best described as a "trust us, we know what we're doing" attitude. It is an attitude that can still be seen with other industries. Due to the large accidents in the 1970s society realised that the chemical industry was potentially a very dangerous one. So they demanded that the chemical industry would open up and show the public what they were doing: "chemical industry, tell us what you are doing". This resulted in a much better relationship between public and industry and also resulted in the Seveso directive. The main aim of the Seveso directive was information to the authorities and the public about what was going on in dangerous sites.
Having received and reviewed the first safety cases, authorities realised that the burden of proof was put on their shoulders. They had to assess the reports and decide if companies were safe. This led in the 1990s to a change of culture. It was not enough that the companies were reporting what they were doing, they had to show that they were doing it in a safe way. This led to the Seveso II directive.

The Seveso directive was fundamentally reviewed and updated resulting in the Seveso II directive (96/82/EC) of December 1996. The changes in the directive are extensive, much more extensive then one would think after casually looking at it. The Seveso directive was basically a "tell us" directive, meaning that companies had to explain what they were doing. The burden was then put onto the shoulders of the regulator to assess the safety situation and make improvement demands. The Seveso II directive is a "prove to us" directive, meaning that companies have to demonstrate that they are safe. This is done without setting the criteria of what is safe and what is not safe. This putş the regulator in a different position. The regulator has to check if the company has fulfilled its obligation and the regulator must decide if it agrees with the safety level set by the company. In addition, there is a large role for the regulator in supervising the companies. This means that the regulators are drawn into a role where they are involved in maintaining the safety of companies. This last aspect is very interesting because it is contrary to the trend in some EU countries and certainly in the Netherlands where government is creating more and more distance between the companies and the regulator. Own responsibility, core business, calculating society are the key words that are used in describing that trend.

However, both the recent disasters in the Netherlands in Enschede (fireworks storage) and Volendam (fire in a café) have led to grave public concern about the role of government in the enforcement of regulations with respect to dangerous situations. In a modern and civilised society the public expects to be protected from all the dangers that threaten life and health. "Acts of god" are not easily accepted anymore. It is very interesting what the result of these discussions will be. General public is not interested in goal setting versus prescriptive regulation, the only thing that counts for them is law enforcement. This debate will almost certainly have a political consequences. It affects all the layers of government in Dutch society. Permits are almost always issued by local authority; at the end of the day the city council and the mayor decides what is happening in the local community. Any change in law enforcement at local, provincial or central government level has to deal with that first. When Seveso II was implemented in the Netherlands the combined ministries tried to raise that issue, in the sense that they wanted to start a discussion about whether, from a technical expertise point of view, it might be preferable to transfer the permit authority from local government to the provinces. At that time that was totally not negotiable. Now after Enschede and Volendam there is no doubt in my mind that the issue will be raised again, and in fact has become reality. The new fireworks directive now moves the responsibility for issuing the permit from the local authority to the provinces.

It is my opinion that it is not very important what sort of legal system one uses in a society, as long as there is a sound regulation and proper law enforcement. However when major hazards are involved government has a direct responsibility and they can not walk away from it by using certification schemes. Again Enschede and Volendam prove these points, the moment there is a large public outcry ministers have to explain to parliament how these things could happen and what they will do to prevent these disasters in the future. Using certification schemes as an excuse is then just not good enough.

The Seveso II directive deals with the dangers for humans and the environment, rather then just the dangers for human beings, as was the case with the Seveso directive.

The directive sets high standards particular with respect to the Safety Management System. I do not think that there is any other European directive that contains as much detail.

Although the Seveso II directive is generally described as a goal setting directive, the direction on how to reach those goals is rather detailed and prescriptive. This was done to ensure consistency in implementation and enforcement throughout Europe. To ensure further consistency, Technical Working Groups (TWG) were established. In these TWGs experts of both government and industry work closely together to provide guidance on some of the subjects of the directive. These subjects are as follows.

1. Inspection systems: this deals with law enforcement by member states. There is an exchange of inspection procedures between the member states.
2. Safety Report: This gives guidance about the content of the safety report. It should contain all the information to convince the authorities concerned that the establishment is safe. This means that depending on the history and safety philosophy of a member state , the information asked can vary considerably between member states.
3. Safety Management Systems. In this guidance the details about what the directive asks for in SMS are explained.
4. Land use planning: Article 12 of the directive states that member states are obliged to develop a policy with respect to land use planning and major hazards. The directive does not give any information on how this should be done.
5. Harmonised criteria article 9: this deals with the derogation criteria which excludes certain substances and/or parts of the site to be taken account of in the safety report.
6. Substances dangerous for the environment: this guidance sets threshold quantities for these substances.
7. Carcinogens: this TWG evaluates the list of named carcinogens in the directive.

About half of these TWGs have drafted guidance on these subjects. The aim of these TWGs is not harmonisation but to capture best practices from regulators and industry all over Europe. The participation of industry in these groups has been extremely useful. Not only were they able to provide the TWGs with very useful examples of working systems, they also seemed to be able to bridge the differences that existed between the various national representatives. The general idea was that if in these TWGs both industry representatives and regulators were able to leave their political "hat" at home they would be able to share their best practices. Again I do not know of any other European directive where there has been such an elaborate effort to draft guidance to help both industry and member states.

There are big differences between the member states. To begin with some member states are federations and some are not. Although Seveso II is an environmental directive, in many countries the Ministry of the environment is not the co-ordinating ministry. In many countries the Labour Inspectorate is not involved. The number of involved authorities per country can vary from two up to five. Types of legislation by which the directive is implemented vary. There is a big difference in culture, know how etc. The sum total of this is that major hazard policy within the member states differs hugely. It is therefore an idle thought that it is possible to harmonise EU legislation to an extent and a detail that within all member states all the companies are faced with exactly the same demands. However there is consensus within the EU about the goal of major hazard policy. In my opinion the way the Seveso II directive was drawn up in combination with the TWGs is the maximum achievable with respect to the harmonisation and detail of EU major hazard regulation.

In this context one should keep the principle of subsidiarity in mind. What is subsidiarity? The EU treaty (article 3b) describes it as follows:

"In areas which do not fall within its exclusive competence, the Community shall take action, in accordance with the principle of subsidiarity, only if and in so far as the objectives of the proposed action cannot be sufficiently achieved by the Member States and can therefore, by reason of the scale of effects of the proposed action, be better achieved by the Community."
EU goal-setting regulation which permits the continued use of good national practice seems the be the best solution to adhere to the principle of subsidiarity. The EU guidance as drawn up within the TWGs can help those countries which do not have good national practice.

As mentioned, the directive is totally revised and many new areas are now covered, to such an extent that it forces member states to develop and investigate their own policies to be able to properly implement the directive. To mention a few topics new in this directive: domino effects, emergency plans, land use planning, information to the public, prohibition of use of the installation, environmental hazards. The scope has been broadened: major hazard policy and safety management systems are now included, the safety report is now used as a tool via which companies have to demonstrate that they are safe, there are time constraints to which companies and regulators have to adhere, and regulators have been given specific enforcement roles with respect to assessment and inspections.
Even the most advanced member states now have difficulties in properly implementing this directive.

THE DUTCH IMPLEMENTATION OF THE SEVESO II DIRECTIVE

In the early phase of the revision of the directive it was recognised by the Dutch authorities that this directive would not be easy to implement. On the one hand there was the local Dutch political discussion about less administrative burden and less governmental involvement with safety issues, and on the other hand, the EU directive pointed in exactly the opposite direction. It was decided between the policy makers of the ministries involved that this directive was going to be implemented in a unique way. It was going to be implemented by one integrated decree, based on four laws, and signed by three ministers. There was political support for this in the form of a motion adopted by parliament which asked the ministers to integrate external environmental safety regulations with internal labour regulations to lighten the burden for the companies. Unfortunately this implied that the burden on the law enforcement authorities would be much higher due to administrative co-ordination problems.

The three ministries involved developed new policies and regulations which all became part of the single implementation decree. The Ministry of the Environment developed a whole new decree with respect to the quality of the environment in order to be able to implement the land use planning demands in the directive, the Ministry of Social Affairs developed a whole new integrated holistic safety approach to be able to properly assess the management system in combination with the technical safety measures and the Ministry of the Interior developed new policy with respect to emergency planning.

New enforcement policy was developed and the three involved law enforcement authorities, Labour Inspectorate, the environmental Permit Authority and the Fire Brigade were given guidance and were sent to integrated training courses to create the best conditions to be able to work together. All in all these are exciting times for all the parties involved. In this process employers, employees and inspectorate were given the chance to give their input to the policy makers.
In this respect the following committees were established: A Seveso plenary committee which consisted of policy makers of all the relevant ministries and their lawyers. Coupled to this were several working parties. In one working party lawyers drew up the necessary regulation. In the second working party policy makers and consultants worked on a Seveso II guidance document for industry and regulators. In the third working party a document was drawn up which described the procedures for the collaborating inspectorates. In a fourth group the results of test sites were discussed. Last but not least, within one project group all the results were extensively discussed with industry, unions and inspectorates. The whole process took more then three years. And all the projects were on parallel time scales. The effect of this is that although the implementation is finished, the result is not perfect. A little bit of debugging is necessary.

Who is leading? The ministries have forced the Permit Authorities, the Labour Inspectorate and the Fire Brigade to work together. They have drawn up an integrated decree based on several laws. The environment ministry is the co-ordinating ministry, which does not mean it is leading. Every ministry keeps its own legal and policy responsibility. The same goes for Labour Inspectorate, Permit Authority and Fire Brigade. Although the Permit Authority is the co-ordinator, every partner keeps its own enforcement responsibility. One of the big problems for companies has always been the contradictory demands of the various authorities. With this new regulation that is minimised. There is one mouthpiece, but every authority has its own enforcement responsibility.

The first results look very promising. Discussions with companies are now integrated discussions and deal with the important issues. The authorities pool together their enforcement powers and are able to operate much more effectively and efficiently. The companies are happier, because the process of being regulated has become much more transparent.

ENFORCEMENT OF THE SEVESO II DIRECTIVE

Labour Inspectorate, Permit Authority and Fire Brigade are concerned with the enforcement of the Seveso II directive. Each one of them is organised in a different way, has a different act and a different culture.

As mentioned before, the co-ordinating authority is the permitting authority linked to the environmental law. This is local authority on a community or provincial level. Since this is a permitting system they issue permits before the actual building of the activity. No activity is allowed unless there is a permit. Within the permitting process there are three departments which are strictly separated.

1) The permitting authority. They are responsible for drawing up the permit and negotiating the terms with the company. There is a certain degree of freedom within the environmental law which can be used by the permitting authority to issue a permit tailored to the local circumstances. The permit procedure is a public procedure meaning that people living in the community have a chance to look at both the permit application and the final permit and are allowed to make formal complaints. The Seveso II safety report is dealt with along these lines.

2) The enforcer. They enforce the permit conditions but are functionally separated from the permitting authority. They use a checklist method to compare what is in the permit with what is the actual situation. If they find deviations the situation on the site has to be corrected. They have the power to fine companies and ultimately they can withdraw the permit. Via the implementation of the Seveso II directive they are made responsible for the government inspections.

3) The policy makers. They are mainly concerned with policy with respect to external safety and risk contours. They check the results of the quantitative risk assessment and assess the consequences of the risk contours.

The combination of these three departments should ensure a proper execution of the environmental law. The system is designed in a such a way that all the demands are made beforehand, before the site is established or the installation is built. Every major change after this, that goes beyond the permit conditions has to be reported and will lead to a change in the permit.

With the Labour Inspectorate the situation is totally different. They are centrally organised and part of the ministry. The law by which they operate is a framework law meaning that everything is allowed as long as one operate within the boundaries of the law. This law, the Labour Conditions Act of 1998, which replaced the 1982 Act, is basically a code for a safety and health management system. Attached to the law there are a series of ministerial decrees which can refer to industry codes and standards. In this way the law ensures that always the most up to date codes and standards are used. Most of these codes and standards are highly technical and detailed, so the final result is a goal orientated safety and health management law in combination with very detailed prescriptive technical codes and standards. Many of these codes and standards were drawn up voluntarily by industry. The labour inspectorate enforces in a repressive way. This means that they go on site and inspect/audit the actual situation. They are not involved in the environmental permit procedure. This means that the labour inspectorate can make costly demands on a site that is already in operation. The labour inspectorate enforces the law in a two step model. In the first step they issue a formal warning or a formal demand. This is usually linked to a time period by which the company has to abide. In the second step they take companies to court. Under the Seveso II directive the Labour Inspectorate can instantly shut down an installation or a site if in their judgement the company has not implemented enough safety measures. This is then assessed by a judge to check if this is a lawful shutdown or not.

Compared to the past, where the labour inspectorate would try to persuade companies to improve themselves this is a radical change and one has to wait to see if it works out. From a political point of view however, the situation has become much more clear. Responsibility has been located with the companies and there is no government involvement.

With the Fire Brigade the situation is even more complex. Fire fighting in the Netherlands is a complex system where local community fire brigades, which are often voluntary, work together with regional fire brigades and the company's own fire brigade. They do not have a law on the basis of which they can make demands to companies. Instead this is done via the environmental permit. They are a formal advisor to the permitting authorities and can as such advise against the issue of a permit if their demands are not met. Their main objective with the Seveso II directive is that they can get enough information in order to be able to establish a community disaster plan and determine the size of the fire fighting and rescue services. If the company refuses to give the information, the mayor can shut down the company.

So what sort of power balance is there between the regulator and the regulated? The chemical industry in the Netherlands is very powerful and influential (gross around 33 billion euro and 78.000 employees). The safety problems are very complex and it is a high risk industry. On the other hand they are very co-operative and active with programs like "Responsible Care", their own very high safety standards, and a lot of international working committees.

The regulator has the power to shut plants down, but does not do that very often. On top of that regulators are weakened by a lack of specialist knowledge and a division of responsibilities. The companies however are very sensitive to receiving formal letters from regulators. All in all an intervention model based on consensus between regulator and regulated will probably work well and has, as a matter of fact, worked well over the years. However, in times when society wants to have everybody's role clearly defined and the responsibilities of regulator clear as well, this model based on consensus is politically not very viable.

FUTURE DEVELOPMENTS

In the introduction to this paper the developments within industry that led to the development of major hazard policy were described. Looking at the organisational trends within industry it is tempting to speculate about what t he future holds. A picture is slowly emerging of global operating companies that will set their own world wide safety standards in combination with a world wide safety management system, leaving a bit of leeway for local regulations and circumstances. Growing complexity of both processes and organisations, global companies with independent business units, corporate goal setting policies with local implementation, outsourcing, increased involvement of the public, these are all trends that policy makers have to reckon with. Goal setting regulation will become more and more important. However the challenge will be to develop the most effective system of (harmonised) codes and standards, private or public, that will give employers and workers enough freedom to still do the job.

Regulators will react to these trends. In this paper I described how regulation developed from "tell us" to "demonstrate to us you are safe". In the Seveso II directive one can already see a large involvement of the regulator with the companies. That involvement will become larger. In my opinion regulation will develop from "demonstrate to us" to "involve us". This is certainly the case in the Netherlands where the permit system in combination with the involvement of the public and the co-operation of the regulators forms a strong basis to build on. Both the recent disasters in Enschede and Volendam and the political reaction show an extremely critical society that wants to be involved in the decision making with respect to dangerous activities. For fireworks storage companies this is too late. Legislation is in preparation which basically could end the fireworks industry in the Netherlands. With cafés and discos, owners have started guided tours through the facilities to show the public that they are safe.

In the next chapter, by Hovden, we see regulators in another context also struggling to define their role in an evolving regulatory system. Hovden's paper is a sobering one. The Norwegian system of internal control, especially as it applied to the off-shore oil industry, had been regarded from its inception as a model for others to follow. Hovden argues that this reputation was deserved in the 1980s but that in the 1990s safety was increasingly sacrificed as a result of relentless cost cutting in the industry. The paper reminds us, therefore, just how fragile are our systems of regulatory control and how dependent they are on economic circumstances. One especially depressing point he makes is that because oil and gas distribution networks have been diversified, the failure of one platform is no longer so critical for customer supply, and safety on platforms may have deteriorated as a consequence. Hovden argues persuasively that safety depends more on the political and economic context than it does on the technicalities of the regulatory regime, and his description reminds us that judgements about safety regimes are always time dependent and subject to change.

Hovden notes an interesting development in new regulations for the Norwegian offshore petroleum industry, namely, a move back to technical prescription. In some matters it will no longer be left to operators to make their own calculations about whether the cost of a protective measure is justified in the light of the risk; they will simply be required to install it. The following paper by Hale, Goossens and Poel describes some dissatisfaction in the Dutch offshore oil industry about the extent of the withdrawal of the inspectorate from detailed rule making and it will be interesting to see if this results in the reintroduction of some elements of technical prescription as is happening in Norway. One wonders, too, whether this trend will eventually emerge in some of the other industries, such as rail, which have moved away from technical prescription in recent years (see Maidment and Becker in this volume).

5

THE DEVELOPMENT OF NEW SAFETY REGULATIONS IN THE NORWEGIAN OIL AND GAS INDUSTRY

Jan Hovden

1. BACKGROUND AND SCOPE

A research project initiated by the Norwegian Research Council has studied the effects on safety and health of new economic, organisational and technological conditions for the exploration and production in the Norwegian petroleum industry (Hovden et al, 2000; Hovden & Steiro, 2000a, Hovden & Steiro, 2000b). The scope of the project also encompassed political implications of the change processes.

Collaboration between the main parties started in 1993 on making the petroleum industry more cost-effective and competitive, in order to face the challenges from big fluctuations in oil prices (10 – 30 $ per barrel), and the internationalisation of the business. The resulting "NORSOK[1]" concept was based on principles of value adding, new types of contracts and teamwork integrating the actors, standardisation, new types of technology, etc. This deal between the government and the industry also stated that the established low accident risk level should be maintained or improved throughout these complex change processes.

---

[1] NORSOK is the short name for "Norwegian Offshore Cost Effective Initiative". See http://www.nts.no/norsok/issued

The regulatory regime of the industry has been through three phases, and we are now in transition to a fourth, as yet unclear phase. The first three phases are as follows.

1. The 70s started with "A Wild West Texan" approach to safety, followed by a modest Norwegiansation and a regulatory regime of the industry based partly on traditions from onshore industry (working environment) and partly on ideas from the advanced nuclear and chemical industries regarding risk analysis, systems safety and management systems for risk assessment and control.
2. The 80s: The Norwegian Petroleum Directorate (NPD) established a strong position and the necessary power to enforce new risk and performance based criteria, e.g. the 10-4 criterion for loosing a platform, internal control of safety and health management systems, etc. A main power base was the control of the assignment of new fields for exploration to the oil companies. Their documentation of safety performance was a main criterion for goodwill. The profit of the industry was good and there were few objections on paying for safety measures.
3. The 90s: Reduced profit and increased competition resulted in the NORSOK concept which initiated fast change in technology[2] and the organising of the industry, including development of numerous standards for co-ordination and control by the industry itself, and an increase in development projects and production, - until 1999. As a response to the dramatic fall in oil prices at the start of that year, the oil companies suddenly stopped all new projects. Even though the prices more than doubled during the year they still act as if the price is 10 $ per barrel.
   The power base of NPD vanished, as controlling the assignment of new development fields became less important. The new fields were less prosperous. In addition NPD was committed by the NORSOK agreement on the changes in the industry. Its strength had diminished throughout the 90s because of fewer resources and because their regulations were designed for the context and organisation of the industry in the 80s and not for a hostile competitive environment of the 90s.

The safety management systems developed in the 1980s were founded on four historical conditions (Ryggvik, 1999):

- the moral authority of the Norwegian Petroleum Directorate (NPD)
- the strong position of the trade unions
- the Working Environment Law (1977)
- the implementation of the internal control of safety, health and environment regulations (1986) specially the emphasis on the company's responsibility and liability regarding safety problems (Hovden, 1998a)

---

[2] From large firm platforms to mainly sub-sea solutions combined with floating installations. As the regulations were mainly risk based, the new technologies were approved by the results of concept risk analyses.

NPD's mandate for control of safety had grown to 16 different regulations. In 1998 they started a process on developing a new regulatory regime and reducing the number of regulations to four or five. An ambition for this process was to win back the initiative in regulating safety.

2. Research Questions – Problems to be Discussed

These changes in the oil industry in the 90s introduce increased uncertainties in defining the risk picture and risk levels and challenge the established routines for risk assessment and control. A main issue has been the role and quality of risk analysis in a setting of new technologies lacking relevant empirical data bases, changes in regulations and hostile market conditions, organisational and inter-organisational dynamics, and the uncertainties introduced by time pressure and parallel activities, i.e. dealing with distributed decision-making. The adequacy of risk analysis for defining risk levels has also been challenged. This article will elaborate and discuss the adaptation of the new regulations to a changing context and climate for risk assessment and control. This includes an analysis of conflict dimensions and the positions of main stakeholders. Some of these are as follows:

A hot political issue in the draft of the new regulations is the paragraphs on reintroducing some basic deterministic safety principles, i.e. barriers, redundancies, "defence in depth" principles, beyond the needs revealed by quantitative risk analysis results of overall risk levels, i.e. precautionary principles for dealing with increased uncertainties. Such recommendations are a threat to short-term profit making and a controversial paradigm for economists and risk analysts. Also proposals on specific safety requirements for high-risk groups and not just population risks for an installation, and requirements on safety culture were strongly objected by the industry.

Measures to reduce uncertainties and improve co-ordination are on the agenda for the new regulations. Means such as applying the new work process orientation of the coming ISO/FDIS 9004:2000(E) management standard[3] for the main structure of the regulations will be discussed. This satisfies the industry's demands on regulations matching the new trends in organising the industry, e.g. BPR, partnership, alliances and multi-organisational teams.

Hot political issues of indirect interest for the subject of the development of new safety regulations, will be excluded, e.g. issues regarding privatisation, internationalisation of contract bids, and regional politics on localisation of supply bases, etc.

However, the new regulations and the reactions to them will be discussed in a framework of contextual and situational factors affecting the approaches and principles chosen. Of special

[3] Main changes from the use of ISO 9004 are stated in the eight main principles for management: a) Customer focus, b) Leadership, c) Participation, d) Process orientation, e) System approach, f) Continuos improvements, g) Fact based decisions, h) Supplier collaboration. The greatest differences are b, c and d. Procedures are to a great extent substituted by the concept of "processes". The process model (PDCA) – similar to the Deming circle, consists of four main processes: 1) management responsibility, 2) resource management, 3) realisation of products and services, and 4) measuring, analysing and improvements.

importance for the regulatory regime, are the effects on safety from the processes of cost cutting. Do the new regulations match the context they will function within, and how adequate are they for controlling the activities and risks of that particular industry?

3. THEORETICAL FRAMEWORK AND EMPIRICAL APPROACH

The theoretical framework for this study was very much inspired by models and discussions in the article "Risk Management in a Dynamic Society: a modelling problem" (Rasmussen, 1997) and by the Bad Homburg workshop resulting in the book "Safety Management: The Challenge of Change" (Hale & Baram, 1998).

The study was based on reviews of documents and data made available by the main stakeholders, and on 30 in-depth interviews with key personal representing all parties and their safety experts. These data was supplemented by three case studies of specific installations. The analyses have been based on qualitative approaches to the study of documents, interviews and cases. In addition, the findings discussed in this chapter will draw heavily on information received when the author was an adviser for NPD in their development of new regulations. An information source added specially for this article, are the written responses from the main stakeholders in the hearing of the draft for new safety regulations.

The study revealed a great number of general trends in economy and market, politics, business organisations and technology of significance for accident risks. These are briefly listed in the box below. These trends are not influencing the accident risks only in a negative way: for some of the trends the effects on safety are mixed, for others we can assume some positive and some negative effects. The main problem is the lack of knowledge regarding the interactions of all these trends.

Economic trends:
- oil prices, financial pressure and market uncertainties
- cost cuts
- economic incentives in contracts
- extreme pendulum movements in investments – from boom to full stop
- internationalisation
- increased uncertainty regarding scenarios for the future

Political trends:
- de-regulation
- pressure on acceptance criteria regarding safety and health
- economic optimisation on the top of the political agenda
- environmental arguments sometimes correspond to economic arguments regarding technical solutions at the sacrifice of safety, e.g. loads, use of materials and substances, physical barriers, etc.
- demands for more cost-effective safety administration and less governmental control
- international commitments by the oil companies, benchmarking, and complaints about expensive and unfair competition

Organisational trends:
- new types of profit/risk sharing contracts making responsibility for safety less transparent, e.g. from client to contractor contracts to integrated teams
- alliances and inter-organisational networks, outsourcing
- parallel activities and distributed decision making
- cuts in workforces, and unmanned installations
- globalisation – international standards

Technological trends and safety challenges:
- more exploration in deep seas and increased volume of sub-sea solutions
- more wells – more blow-outs
- deep sea production with flexible raisers
- increased number of floating installations
- more simultaneous operation and modification
- more oil and gas activities in the hostile waters further north in the Norwegian sea
- problems regarding the disposal of scrapped platforms

4. THE COST CUTTING CHANGE PROCESSES

The "architect" behind the NORSOK concept was the Minister of Oil and Energy, and it was based on a three-party co-operation, although one of the two main unions was reluctant to participate. The other one was also critical, but accepted that a major restructuring was necessary for the sake of employment and the competitiveness of the industry in a global dynamic market. The NORSOK initiative was also followed with improved conditions for the industry regarding taxation and opportunities for international actors. After a period of discoveries of new oil and gas sources and increased production rates the situation in 1993 was characterised by:

- decreasing oil price
- flattening of production volume in the North Sea area and increased costs per unit
- a need for upgrading of older installations
- smaller new discoveries, and disappointing results of exploration in the Barents Sea

The established NORSOK forum developed seven priority areas:

1. Improved cost analysis
2. Standardisation (design principles, common requirements, specific system requirements) - about 100 standards developed so far
3. New relations and contracts between operator and suppliers/contractors including concepts for new work processes
4. Documentation and information technology
5. Restructuring base and transportation activities
6. More cost-effective safety, heath and environment management
7. Framework agreements between government and industry stimulating increased activities

The NORSOK initiative formulated two main goals:

- A 40-50 % reduction in costs and time within five years
- Maintenance of the established safety level and, by benchmarking, maintenance of the Norwegian oil industry as one of the safest in the world

Although the NORSOK concept covers the entire lifecycle the main impact has been on the development and construction phases, see figure 1 below. This new approach created boom activity in the new development projects and in the construction and supply industry activities far beyond their normal capacity. This was followed by a panic reaction the winter 1998/99 resulting in a full stop as a response to a short term oil price of less than $ 10 per barrel. The price grew fast to $20-30, but in a power-play with the authorities for improved conditions, i.e. reduced safety requirements, the industry used $10-12 as a reference price on oil.

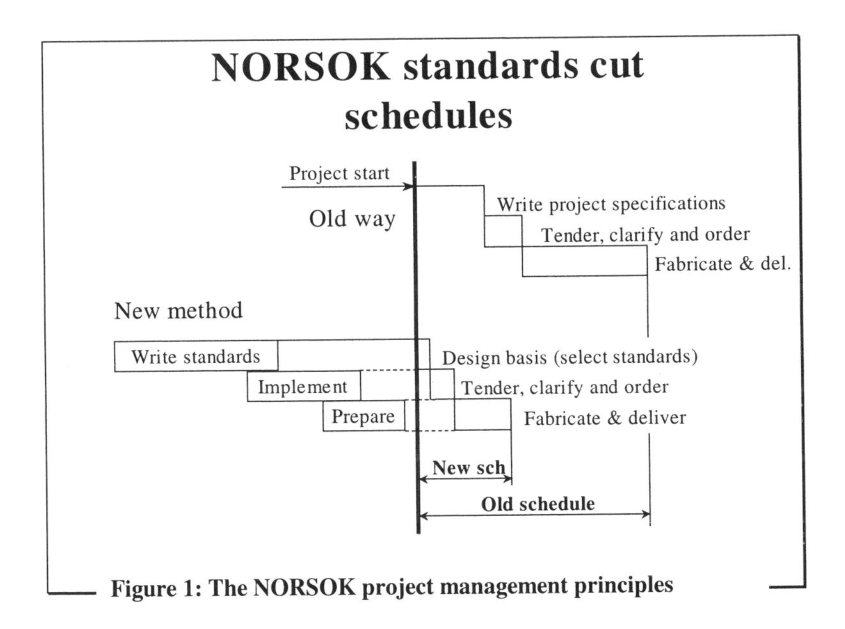

Figure 1: The NORSOK project management principles

An allegory on the "meaning" of NORSOK (told by an oil company employee), and partly based on experiences from contracting floating installations from the Far East:
"Earlier we bought a Mercedes for the sake of quality and safety. Now we are told to buy a cheaper Volvo, which gives the same safety level. The cost cutting requirements and budget force us to buy a Lada camouflaged as a Volvo to satisfy top management. Then we have to go through a costly upgrading and reconstruction of the Lada to meet the specifications of a Volvo."

Many NORSOK based development projects experienced huge budget overruns. However, they were still far cheaper per unit of production compared to installations built in the earlier period. The NORSOK secretariat ascribed 60-70 % of the savings to technological improvements and innovations; the rest is related to new contracts and improved work processes including the use of NORSOK standards. For the operating installations the NORSOK effects have been more modest so far. However, it has resulted in more pressure to reduce manning, outsource functions, e.g. maintenance, and to upgrade technology. The recent developments in ICT have given opportunities to move control room functions onshore.

NORSOK has more and more been used as a common term for *all* change processes in the industry and not just those defined by the NORSOK secretariat, i.e. the seven priority areas listed above.

5. The Risk Picture: A Mixture of Facts, Uncertainty and Myths

According to studies by Vinnem (1998) and Hauge et al (1999) the following trends in the risk picture for offshore installations can be identified:

- an increase in major hazards – mostly related to problems of gas leakage (loss of containment),
- an increased risk and uncertainty regarding safety levels for floating installations, - mostly related to new advanced and untried technologies,
- an increased exposure to the high-risk arena of shuttle traffic by helicopters,
- an increased accident risk related to the activities of service vessels,
- an increased vulnerability caused by increased use of ICT systems, automation, and reduced manning on the installations,
- low and stable frequencies of occupational accidents (LTI-rates) for the last decade.

The representatives of the oil companies found comfort in the low LTI rates as indicators of successful accident control. The safety information system applied by the whole industry (SYNERGI) is based on the ILCI model and the underlying "iceberg" theory (Bird & Germain, 1985; Heinrich, 1931). However, some incidents, especially gas leakage, with a catastrophic potential have occurred in the last five years despite excellent LTI rates. Such incidents challenge the widespread belief in the "iceberg" model for the detection of major hazards by empirical analysis of near-misses and occupational accidents, and made some of our respondents ask for alternative analytical approaches for the study of hazardous processes in dynamic complex systems. The strong belief in common causes of minor and major accidents is a group-think phenomena in safety management in many industries and in safety research based on misinterpretations of Heinrich and other classical writers in the field (Hale, 2000). In his extensive review and discussions of the subject A.R. Hale names this fallacy "an urban myth".

On stationary installations the increase in fatalities have been marginal since 1993, whereas the increase of fatalities on drilling rigs has been significant, and on stand-by vessels there has been an even greater increase in the fatal accident rate. Helicopter accidents are still the dominant risk factor regarding human lives. Due to changes in technology the high-risk activities of diving have disappeared. Serious violations of safety and health regulations discovered by the NPD in the last years resulting in closing down the business and penalties have been dominated by drilling rigs, especially British owned, with scandalous working environment conditions. A company with traditionally a high safety profile, Norsk Hydro, lost its safety reputation when serious and systematic safety omissions were discovered during the investigation of a fatal accident.

A historical study of the Norwegian offshore safety regime by Ryggvik (1999) concludes that from being an ideal and a model for other industries nationally and internationally in the late eighties, the safety systems have crumbled during the nineties. A review of the offshore emergency preparedness (Carlsen and Schrøder, 2000) concludes that emergency response capability has been reduced.

Confidence and conceit based on historical records and successful priorities of safety measures in the past may currently represent the greatest threat regarding major hazards in the industry. When the oil industry began acting high and mighty on safety issues the reactions from NPD, the reactions of the unions and the public were reduced trust and confidence in the industry's safety priorities.

So far the established standards for safety performance have been "sacred", but in a time of empty order books and downsizing it is easy to predict a pressure on reduced safety and health standards to "save" the industry and workplaces. Such arguments have become part of the hidden agenda between the parties, but so far very little visible in the public debate. Moreover they abated when the oil prices rose again.

We have found no clear support for the complaints that formal safety requirements in the Norwegian oil industry give a competitive disadvantage in the international arena. Possible differences may be due to a more proactive and effective control regime (NPD) and the traditional high priority placed on safety by the oil companies themselves.

6. The Development of New Safety Regulations

Currently, NPD is in a process of rewriting all their regulations. This is an important arena for questions regarding types of control regime and resources, priorities and principles regarding risk assessment and control.

The process of developing new safety regulations by NPD started in 1998 when internal working groups were appointed. Very soon these NPD work groups established reference groups to support and correct their work. The reference groups represented a mixture of experts from R&D institutions, the oil companies and their federation, the unions, and more, i.e. the participative aspect of representatives of different stakeholders was established at an early stage. In addition the work groups looked for ideas for regulatory concepts from different countries and industrial domains, e.g. the nuclear industry. The draft regulations were on an official hearing among the stakeholders from June to November 2000.

The regulatory safety regime of NPD had lost the initiative in the NORSOK process of cost cutting, which they as part of the political system was expected to loyally support. Economy and oil prices had for some years overshadowed safety issues on the political agenda. The aims of developing a new regulatory system were to replace outdated regulations, and to win back more power for enforcement and the initiative in regulating safety in a proactive manner.
The internal control regulation of SHE and all the 15 specific regulations regarding risk analysis, emergency preparedness, working environment, etc. have been replaced by a framework regulation, and four more detailed pieces of regulation on:

- management of safety (SHE)
- operations (activities)
- technology (installations)
- documentation (information)

The safety management regulation also includes the requirements derived from the general pollution and the health regulatory authorities. Their control authority is delegated to NPD. The new safety management regulation follows two lines or strategies:

1. Adjustment and adaptation to demands from the industry:

- simplification, reducing the number of regulations
- definitions and vocabulary based on international standards
- a structure and logic of the contents plus requirements based on the coming ISO 9000 series (ISO/DC2 9001:2000, ISO/DC 9004:2000), i.e. adjusted to other management systems applied by the industry and the revealed need of changing focus from auditing systems to auditing processes and resources
- the main principle for formulating acceptance criteria is risk informed functional criteria

2. Counter-pressure and suggestions challenging the industry

- Especially chapter 2 “Basic requirements and principles” overrules functional risk based criteria by formulating some absolute deterministic requirements regarding safety barriers and redundancies. Also new are requirements related to safety culture, originally in the safety management regulation but moved to the framework regulation (§ 22), as a way of focusing on management competence and commitment without strong reliance on bureaucratic systems and documentation.
- In paragraphs on risk acceptance specific (small) high-risk groups are mentioned, not just the overall FAR value for an installation (population risk)
- More articulated requirements to the management of (parallel) activities, processes, ad hoc organisations, manning and competence based on the changes in the industry the last decade

The contents of the new safety management regulation[4]:

Chapter 1 *Introduction*

1. Definitions

Chapter 2 *Basic principles and requirements*

2. Principle SHE management requirements

3. Qualification and control of other actors in the industry

4. Principles for risk reduction

5. Barriers

Chapter 3 *Goals and decision criteria*

6. Goals and strategies for safety, health and environment

7. Decision processes

8. Establishing specific requirements

9. Risk acceptance criteria

10. Performance measures and indicators

11. Principles for safety systems

Chapter 4 *Accomplishing petroleum activities*

12. Management of work processes

13. Resource and activity planning

14. Deviation control

15. Investigation of accidents and near-misses

Chapter 5 *Management of resources in the petroleum activity*

16. Manning and competence

17. Collection and distribution of data

18. Co-ordination of analyses

19. Analyses to reveal risks of losses and pollution

20. Use of results from analyses

21. Updating

22. Analyses for identifying major hazards

23. Quantitative risk and emergency analyses

24. Environmental risk analyses

25. Working environment analyses

Chapter 6 *Supervision and improvements*

26. Internal supervision and auditing

27. Verification and validation

28. Development and improvements

[4] Draft of June 2000 for the formal hearing of the parties

Each of the paragraphs are generic and with little text, just a few lines, followed by more detailed comments and guideline with hints, examples, etc. in order to make the understanding of the intentions of the regulation clearer.

A special problem not fully solved is the interface and "grey zones" between maritime activities and petroleum activities and on crossing the borders in the North Sea. Some compromises and co-ordination have been made through the North Sea Offshore Authorities' Forum (NSOAF). NSOAF decided that the differences in safety regulations and practice among the safety authorities of the five North Sea countries are so small that a common market for floating installations based on an approval procedure/certificate of acceptance will be established within the end of year 2000 (Hagland, 2000). Drilling rigs have been moved around the world for years but have had to adjust to quite different regulatory regimes. As floating production, storage and offloading (FPSO) installations/ships replace the stationary production platforms, the earning potential increases considerably with prospects of moving between oil and gas field in the whole North Sea area without regulatory constraints.

7. FINDINGS: THE ROLES AND POSITIONS OF THE MAIN STAKEHOLDERS

The main stakeholders interviewed in the study were:

- the regulatory authority, plus the ministry responsible for offshore safety
- the oil companies
- trade unions
- suppliers and contractors, incl. construction companies
- the trade organisations for oil companies and for the contractors
- engineering companies and consultants

The interviews reveal an improved interplay between the oil companies and the contractors in development projects on introducing safety and health issues at early stages and in solving safety problems. Our respondents were quite critical of the NORSOK technical risk analysis standards, but everyone praised the working environment standard as a success.
New technology has been the dominant change factor both regarding cost cuts and safety challenges. Cuts in workforces and manning are an integral part of the design criteria and total risk assessments for new installations. On existing installations they try to reorganise the work processes and reduce the manning far below what the installations were designed for. This creates very difficult decisions and involving affected employees has been crucial for a successful outcome (Hvalgard, 1999).

Some of the main conflict dimensions revealed by the interviews were:

Safety versus cost

Officially such conflicts were a taboo subject, i.e. the NORSOK agreement on maintaining high safety levels, but they were nevertheless widely discussed. A lot of anecdotal information was given on cost cutting affecting safety, from the unions but also from safety experts within the oil companies. Unofficially the Norwegian SHE requirements was attacked – especially from the trade organisation of the contractors, for destroying the competitiveness of the industry. The two main unions revealed different positions. One was strongly opposed and non-co-operative to the NORSOK agreement based on the threats to the workers health and safety. The other was co-operative due the needs for competitiveness and for future employment in the industry, but ambiguous and critical to the practice of the agreement.
Safety and emergency aspects of down-seizing on old installations worried NPD and the trade unions more then the others. Means for improved change management encompassing safety were much discussed.
The traditional safety management systems were seen as extremely formal, rigid, and with detailed requirements concerning documentation. This bureaucratic documentation requirement was strongly criticized for its effect on costs, both by line management in the oil industry and the contractors. This attack was also directed at external safety consultants, who were blamed for driving up costs.
The stakeholders did not agree on who are the winners and who are the losers of the NORSOK process, but a majority of the respondents seemed to agree on that the Norwegian state had been the great winner but also the oil companies. The results for the main contractors were more disputed, whereas almost everyone agreed that losers were the small sub-contractors and especially the engineers and workers at construction sites had suffered from extreme cost and time pressure.

Oil industry versus maritime tradition regarding regulations

The development of floating installations, e.g. combined drilling and production ships, have brought in shipowners as important actors in addition to their role in the rig market for years. Their regulatory regime is traditionally based on pre-approval certificates, which give predictability compared to the internal control philosophy and functional requirements practised by NPD. A tension and unresolved conflict between the regulatory principles of the maritime sector and the oil sector were observed.

"Model power" and other issues and dilemmas

Differences in perception of risks and mental models of reality were striking. Representatives of the industry referring to accident statistics stated that "safety has never been better", but this was not agreed on by NPD based on their knowledge and data, and it was strongly opposed by the experiences and perceptions expressed by the workers. Management and (technical) risk experts believed strongly in the superiority of quantification and numbers for describing an "objective reality". The unions and safety experts with a background in social sciences pointed at the weaknesses and manipulative effects of the "hard data". As a part of this, the focus on LTI rates and the "zero vision" adapted by Statoil last years were highly disputed, but the lines of conflict went across the stakeholders and not between them.

The quality of risk analysis and risk assessments regarding new technologies (floating and sub-sea installations) especially worried NPD and some of the experts from the oil companies more then other stakeholders.

It was also noted that practices which are adopted for environmental reason can also increase the risk of accidents (e.g. re-injecting gas)

The position of risk analysis has changed. Before NORSOK a main purpose was to give decision support to improve safety beyond the minimum requirements and to prove the trustworthiness of installations according to functional performance criteria. Now the search for improvements stops at satisficing solutions within a framework of rather short-term economic optimisation. The risk analyses are less used as decision support and more as a means for verification. This is partly a consequence of the many parallel activities and time pressures in development projects. Respondents from NPD and also some of the safety experts within the oil companies and engineering companies expressed a worry about the quality of the risk analyses in monitoring the complex changes in the risk picture. The combined uncertainties and the lack of transparencies of the socio-technical systems regarding the boundary for safe performance made some of the respondents express a fear of disasters waiting to happen. This made some of our respondents ask for alternative analytical approaches for the study of hazardous processes in dynamic complex systems.

To improve the quality of risk analysis especially in dealing with the operational phases NPD initiated a project (Øien, 2000) not just to obtain more correct risk estimates but to develop new means for risk control. The project has great similarities to I-risk (Oh, J.I.H. et al, 1998) and SAM (Murthy & Paté-Cornell, 1996). Some oil companies were very critical to the project, officially on methodological grounds, but mainly due to fear of cost.

These interviews summarised above describe the context for the regulatory regime. However, they were not linked directly to NPD's development of new regulations, so the answers from the same stakeholders to the hearing of the draft regulations correspond more directly to the aims of this article. The main responses to the draft regulations from four representative stakeholders are as follows:

Norwegian Oil Industry Association (OLF)

The form and contents of the new regulation will be more user friendly, flexible and logical. The formulation of functional requirements with reference to standards is appropriate.
Joint regulation based on many laws and authorities co-ordinated by NPD should be followed up by more harmonised and cost-effective control functions. Due to overlap the safety management regulation should be incorporated in the safety framework regulation. The evaluations of administrative and economic consequences of the regulations are superficial and partly missing. OLF proposes a forum for collaboration with NPD regarding the implementation and updating of the new regulations.
OLF notice that the new regulation focuses more on the responsibility and duty of the owners, but still this is not complete and consistent enough. The required analytical tools for environmental risks are not fully developed. OLF approve the possibility of applying maritime norms.
The regulations should to a large extent promote the development of new technical solutions. However, the drafts make some delimitation regarding new socio-technical concepts and models (i.e. deterministic requirements on barriers, redundancies, and emergency preparedness).

Norwegian Shipowners' Association (RF)

The use of the "SHE" concept introduces a belief in this area that something special is to be performed in addition to other management functions. The safety management regulation should be encompassed in general management guidelines/regulations for the industry. How the SHE requirements are related to other requirements is confusing. Costs increase due to unclear regulations. The scope of the regulations is unclear due to the mixture of laws encompassed: terms like installation, device, vessel, etc. are used in the text without clear references to the scope of the law they are founded on.
RF opposes the abandonment of specific pre-approval for floating installations in the new regulations, i.e. a need for single decisions founded in the regulations.
RF finds too many references to NORSOK standards instead of international standards. The International Safety Management (ISM) code for maritime activities should be accepted as a standard for satisfying the requirements on goals and strategies for SHE.

The Workers' Federation of the Oil Industry (OFS)

There is no need for new regulations. It will just contribute to continued negative SHE performance in the oil industry. The safety regulations are based on confidence between parties, which does not exist any longer. The pillars of openness and trust in a three-party collaboration are crumbling. The understanding of reality regarding the risk picture and safety efforts in the industry differs significantly.

The new regulations contain fewer regulations and pages. But with the new information technology the challenge is to develop regulations designed to reduce to a minimum the needs for external information sources. Founded in the working environment law the regulations should be formulated in a language comprehensible for layman judgement. This is violated by the expert oriented text of the new regulations. Furthermore, the extensive use of references to external standards presupposes access to electronic information systems. This excludes large groups of workers without knowledge and/or access to the electronic information systems. There is a need for popular explanatory texts instead of the sparse text of references to standards. The down-seizing of safety expert staffs in the companies has reduced the number of professional persons able to interpret and follow up the safety regulations, adding another argument against the slim texts.

The oil industry is producing a great number of disabled people. It is ironic that the LTI statistics is presented with many decimals whereas the companies have no statistics at all on workers made invalid. Injuries and diseases related to noise, chemicals, ergonomic loads and strain, shift work, etc. are neglected. The regulations should have requirements to prevent misuse of statistics.

The oil companies squeeze the contractor companies on prices. The contractors are doing the most dangerous work and are squeezing their workers' health and safety. The safety regulations should give more focus to the contractor systems.

More and more important decisions are taken by the board of licensees of a field, and not by the operating company where the workers are represented. The new regulations do not handle that problem.

The interpretation of the use of "shall" in the regulation itself and "should" in the comments to the regulation causes confusion. "Should" is sometimes a recommended but not mandatory way of fulfilling the requirements, at other places "should" means "shall as a minimum" depending on the context of the specific regulation and comments jointly.

Norwegian Oil- and Petrochemical Workers Union (NOPEF)

Regarding the use of "shall" and "should" they have the same comments as the other union. In many other aspects the two unions differ. NOPEF praises NPD for the radical changes in the regulatory regime especially the move from detailed requirements to functional requirements. In the old regime the minimum requirements were regarded by the industry as maximum as well, and for this reason the climate for improvements has not been positive. However, a change to more functional requirements will require more intensive control activity by NPD. NOPEF does not oppose the references to NORSOK and international standards, but it is concerned that the standards might change for the worse without any influence of workers' representatives. They are

critical to the "ALARP" principle in the framework regulation (§ 20), especially the guidance on how to make cost/benefit decisions about risk reduction. .
NOPEF also mention some confusion regarding the responsibility of licensees versus operators on installations. However, generally the new safety management regulation got many words of praise from NOPEF.

8. Discussion: Power and Politics

The discussion following relates to main questions for the book as described by Hopkins & Hale in the Introduction.

A process of evolution and participation, but also conflict

Part of the old regulatory regime encompassed goal oriented functional criteria for major hazards based on risk analysis in the concept and design phases (20 years of practice), and requirements of safety management systems and auditing (15 years of "internal control" practice). On the other side, quite detailed prescriptive rules dominated many of the regulations. A main aim of the new regulations was to transform the prescriptive regulations to external standards. As argued by the union OLF this is no real simplification, and it also creates problems for lay understanding.
It is also interesting to notice that the two stakeholders, OLF and NOPEF, which were strongly involved in the development process responded positive to hearing draft, whereas the two others, RF and OFS, were respectively ignorant and strongly negative. The responses from the oil companies were more critical of the regulations and NPD in interviews than in the hearings response. The reasons can be a combination of: (1) they had some success with their viewpoint in the development process, (2) they changed tactic during year 2000 from being arrogant on safety issues to a co-operative approach due to changes in political climate and public awareness. The safety authorities, supported by research and documentation from NPD, in interplay with media focus on specific safety scandals in the oil industry, succeeded in bringing the industry in a defensive and more humble state.
A serious problem revealed both through the interviews and the hearing responses was the lack of agreement on how to "discover risk and define the problem", i.e. the starting point for the whole problem solving cycle (see fig. 2 in Chapter 1 by Hopkins and Hale). To deal with this problem NPD has established a pilot consensus project on defining and identifying risk levels. The project is manned by risk experts from NPD, the industry, consultants, and researchers working as a cross-disciplinary team involving disciplines from pure statistics to anthropology. The output from the project is the basic information for a consensus forum representing all main parties. The intention is to make the project and the forum permanent for dealing with conflicting perceptions of reality

A regulatory regime striving to match the threats from changing market conditions and the fast pace of technological change (the context of NORSOK processes)

The NORSOK can be considered as a change process for adjusting the costs to other international actors in this industry. It should contribute to more effective use of resources. The evidences revealed in this study regarding the intentions of maintaining a high safety level combined with more cost-effective safety management give little general support for a successful NORSOK process. On one hand, new technology and organisational changes have reduced cost for upholding safety requirements. On the other hand, the increased use of risk analysis for the purpose of economic optimisation has reduced the safety margins and the robustness of the systems to withstand major accidents, and also introduced more uncertainties in assessing the factual risk levels. However, the total production and transportation system is more robust, which may reduce government's and oil companies' motivation for safety at the level of installations.
Reduced vulnerability of the whole system (society, infrastructure and companies) may increase the willingness to accept higher risks at a lower level, i.e. the risks on installations and work-places. The increase in activities, the development of new and to a large extent mobile installations and a number of new pipelines for oil and gas compared to the old situation of a few big installations and almost all transportation by pipelines routed through the Ekofisk tank, has dramatically reduced the *vulnerability* of the infrastructure. It makes the regularity of production and delivery more reliable and robust. Based on pure economic reasoning it can be argued that Norway as a nation and the oil companies now can "afford" a major accident – a new "Bravo" blow-out, a "Piper Alpha" or "Alexander L. Kielland" type of accidents. Of course, they have to add the costs of image loss, political consequences, etc., but nevertheless based on pure economic criteria government and companies may loose motivation for safety on oil installations and reduce safety margins, accepting more gambling and uncertainty. The pressure on safety margins comes from many sources:

- the politicians and the public: safety as a threat to the big national funds based on incomes from the oil activity as a basis for future pensions and for developments in onshore Norway, especially the public sector;
- the unions: safety as a threat to employment in the oil industry and the supply industry – corresponds to the stand of the oil companies and the supply industry: safety as a threat to their competitiveness in an international market, their profit margins, etc.

In the years 1998-99 with great ups and downs in oil prices, the oil industry went from excessive investments and lack of capacity to a complete stop in investments. In this turbulent situation we have observed an intensive political struggle between all the main stakeholders – government, regulators, unions, oil companies, the construction industry, suppliers and contractors. In the year 2000 with high oil prices, the voice of safety issues on the political and public arena became higher and clearer. The results of this struggle are not clear or easy to predict yet. However, conclusion to be drawn from this study is that the success or failure of a regulatory regime is more dependent on the political climate and public awareness than the scientific basis for applying alternative administrative principles of regulation and control.

The independence and integrity of the regulatory regime in dealing with the levels in risk management

The Norwegian State is under conflicting pressures in balancing safety and economy. The Norwegian economy is very dependent on the oil sector and the government has high ambitions for the funds they are building up based on taxation and earnings from the oil business. The state is also an active owner through Statoil and partly Norsk Hydro, as well as a major licensee on many fields. The NPD is divided in two main functions: one for resources, supporting exploration, production and transportation, the other one for safety. They have a common co-ordinating director on the top, reporting to two different ministers - on the resource side to the Oil and Energy Ministry, and on the safety issues, to another ministry with no business interests in the oil sector. In any major clash between safety and economic interests two ministers with different mandates and finally the whole government will have to deal with it. This arrangement has been successful for years, and the British system copied it after the Piper Alpha accident.

Safety awareness and priorities are different at different levels in the oil industry. The operating organisation in the field is responsible for the safety management systems, and they usually take their role seriously according the internal control principles. This organisation is however under strong pressure from the group of licensees (the shareholders), who are asking for benchmarking on productivity and costs per barrel. In that group both their own oil company, other oil companies and representatives of the state ownership do not bother about safety at all. For 15 years there has been a strong focus on management responsibility for safety. The next debate will have to be on the responsibility of shareholders and licensees. The NPD draft regulation gives no answer to that challenge.

The new regulations give more attention to problems of responsibilities in complex contractor systems. However, the changes in organisational constellations happen so fast that it can be questioned if the proposed regulations are able to cope with those control problems.

International market forces are pressing for deregulation, harmonisation and standardisation. NPD's new safety regulation goes a long way in accepting international standards. As mentioned above NPD lost the internal control principle of no pre-qualification in the safety control of rigs to North Sea common market certification. The proposed new regulations are defensive in dealing with how to match the forces of globalisation.

Is the draft regulation an adequate answer to the threats and challenges?

The draft combines both adaptations to the new realities in the industry and countermeasures to maintain a high safety level. The legitimacy of saying *no* based on deterministic or precautionary principles, and defying the results of risk analyses may become important in controversies with the industry especially regarding uncertainties introduced by new technologies (barriers), new organisational concepts (redundancies/manning), etc. The success will depend on the political legitimacy of these means.

The requirements of safety culture will be a challenge for auditing validity, but if successful, it may be an useful supplement to more formal systems approaches.

An important feature of the draft regulations is that it challenges the dominant and "politically correct" way of thinking, i.e. the utility basis of economic rationality and reasoning (Hovden, 1998b), adding:

1. Justice by asking for sub-group risk analysis (the risk exposure for specific small high-risk groups of employees) not just population risk analysis.
2. Discourse by participation and involvement from all relevant parties from the start of the development process to the end.
3. Duty by defining responsibility in different types of organisations including ad hoc organisations and inter-organisational complexes. However, the safety responsibilities for shareholders and licensees are not properly defined.

To draw conclusions about the success of the new regulations, we will have to look at the implementation and the enforcement of the means for control.

REFERENCES

Bird, F.E. & Germain, G.L. (1985) *Practical Loss Control Leadership.* Institute Publishing. ILCI. Georgia.

Hagland, J. (2000) "Et hav uten grenser"("A sea without borders") *Sokkelspeilet (The Shelf Mirror),* no 1/2000

Hale, A.R. (2000) "Conditions of occurance of major and minor accidents". Paper at the seminar *Le risque de défaillance et son contrôle par les individes et les organisations.* Gif sur Yvette.

Hale, A.R. & Baram, M. (1998) *Safety Management. The Challenge of Change.* Pergamon, Elsevier Science Ltd.

Hauge, S., Westby, O., & Jersin, E. (1999) *Trender i petroleumsvirksomheten – hvordan kan de innvirke på risikobildet? (Trends in the petroleum industry – how can they influence the risk picture?)* Trondheim: SINTEF report, SINTEF Industrial Management.

Heinrich, H.W. (1931) *Industrial Accident Prevention.* McGraw Hill. New York.
Hovden, J. (1998a) "Models of organisations versus safety management approaches: A discussion based on the "internal control of SHE" reform in Norway". In Hale, A.R. & Baram (eds.) *Safety Management. The Challenge of Change.* Pergamon, Elsevier Science Ltd.
Hovden, J. (1998b) *Ethics and Safety: "mortal" questions for safety management.* Paper for the conference "Safety in Action", Melbourne, 25-28 February 1998.
Hovden, J., Hvalgard, K., Steiro, T., & Sten, T. (1999) *HMS-utfordringer for norsk petroleumsvirksomhet. (SHE challenges for the Norwegian petroleum industry)* Trondheim: SINTEF memo.
Hovden, J. & Steiro, T. (2000a) "The effects of cost-cutting in the Norwegian petroleum industry" In Cottam, Harvey, Pape, & Tate (eds.) *Foresight and Precaution.* Balkema, Rotterdam.
Hovden, J. & Steiro, T. (2000b) "Distributed decision-making and uncertainty management: Accident risk assessment in petroleum development projects". In Kondo & Furuta (eds.) *PSAM 5. Probabilistic Safety Assessment and management.* Universal Academy Press, Inc., Tokyo.
Hvalgard, K. (1999) *Sikkerhet ved nedbemanning offshore (Safety aspects of down-seizing offshore)* Master thesis, NTNU.
Murphy, D. M. & Paté-Cornell, M. E. (1996) "The SAM framework: Modeling the effects of management factors on human behavior in risk analysis". *Risk Analysis,* pp. 501-15.
Oh, J. I. H. et al (1998) "The I-RISK poroject: Development of an integrated technical and management risk control and monitoring methodology for managing and quantifying on-site and off-site risks". Proc. from *PSAM 4,* pp 2485-91, New York.
Rasmussen, J. (1997) "Risk management in a dynamic society: a modelling problem." *Safety Science,* Vol. **27**, No 2/3, pp. 183-213, Elsevier Science.
Ryggvik, H. (1999) *Fra forbilde til sikkerhetssystem i forvitring. Fremveksten av et norsk sikkerhetsregime i lys av utviklingen på britisk sokkel (A safety system from ideal to crumbling: Comparing the raise of a Norwegian safety regime with the British Shelf)* Memo no 114, TIK-senteret, University of Oslo.
Øien, K. (2000) "A Structure for the Evaluation and Development of Organizational Factor Framework". In Kondo & Furuta (eds.) *PSAM 5. – Probabilistic Safety Assessment and Management.* Universal Academy Press, Inc., Tokyo.
Vinnem, J. E. (1998) *Risk Levels on the Norwegian Continental Shelf.* Stavanger: Preventor, report no 19708-03.

The paper by Hale Goossens and v.d. Poel is a case study in what it has meant for a regulator to move from prescriptive to goal setting regulation. They examine the Dutch offshore oil inspectorate at two points in time, roughly a decade apart, which enables them to identify very clearly the change which has occurred. They argue that the change has not been a retreat from rule making and enforcing, but rather, a change in the kinds of rules which the inspectorate makes and enforces. The authors lay out a taxonomy of rule making and enforcing and they demonstrate just where the inspectorate fits within this total framework of regulatory activity. It is a framework which could usefully be used by others who seek to study the activities of safety inspectorates.

The transition which the Dutch inspectorate has undergone requires considerable re-skilling and this has not always been easy. Again this is a theme echoed in several other papers in this volume.

The study does not aim to determine whether the new regime is more effective than the old in assuring safety, but the authors do report both positive and negative outcomes and many matters still to be resolved. It is interesting, then, that neither this paper nor the previous one give unequivocal endorsement to the move away from detailed prescription.

6

Oil and Gas Industry Regulation: From Detailed Technical Inspection to Assessment of Safety Management

Andrew Hale, Louis Goossens, Ibo v.d. Poel

1. Introduction

1.1. Background to the studies

This chapter arises from a study carried out for the Staatstoezicht op de Mijnen (SodM), the Dutch inspectorate responsible for regulation and inspection of the mining industry. The inspectorate was set up in the 19th century to regulate the Limburg coalmines. After these closed in the 1970s, it moved its activities almost entirely to the regulation of the on- and offshore oil and gas industry. It also retained control of the "aftercare" of the coalmines (e.g. ground subsidence) and the marl and salt extraction, but these are minor activities compared to the oil and gas.

In the aftermath of the publication of the Cullen report (1990) on the Piper Alpha disaster, there was strong pressure to make far-reaching changes in the way in which offshore inspectorates carried out their role of regulation and inspection. One of Cullen's main recommendations was to shift the focus of inspection from the details of technical compliance to the establishment and assessment of safety management systems and safety cases. The Dutch government decided to follow this approach also. Working parties of SodM and industry representatives were set up to consider whether and how to implement Cullen's recommendations in the Dutch sector of the

North Sea and in land-based oil and gas extraction. In particular, SodM decided very early to adopt the safety case and safety management system recommendations. This move was in line with the general trend in government inspection heralded by such reports as the Robens Commission (1972) in the UK, the Internal Control regulations (1985) in Norway and the Working Conditions law in the Netherlands (1980). SodM was in many senses behind this trend in regulation and was still operating as a "hands-on" technical inspectorate.
At about the same time the Algemene Rekenkamer (Dutch Public Accounts Department) carried out an assessment of all of the inspection services of the various ministries. One of their criticisms of SodM was that there was no external and independent audit of their work and its effectiveness. In 1990 one of the main unions (FNV), with significant membership in the oil and gas industry, also put in a complaint to the Ombudsman about the working of SodM. It alleged that inspectors were not independent of the operating companies, that they failed to communicate enough with the union and that they did not carry out their controlling task effectively.

As part of the response to all of these pressures, but particularly of the first two[5], SodM commissioned the Safety Science Group in 1991 to conduct a study of the work of the inspectorate and the way it was perceived by all parties and to make recommendations for change. The next section summarises the study and its findings. As a result of the study, the Group was asked to set up a training programme to equip all of the staff for the new role and working methods. This programme was carried out in 1992-3 and is also briefly described in the next section. As a result of these interventions, and of its own initiatives based directly on the Cullen report, the inspectorate began the process of change of regulatory philosophy and of organisational structure and functioning.

In late 1999 the Safety Science Group decided, as part of a faculty research programme on the effect of regulation on technological development, to revisit its work with SodM to see whether the recommendations for change made in 1991 had been effective. Data were collected from the study of documents over the period from 1991-2000 and a series of interviews were held in 2000 with inspectors and with representatives from the oil and gas industry about the changes which had occurred in the ten years since the first study.

This chapter discusses a range of issues concerning the role of regulators in the oil and gas industry and the problems encountered in making the change from detailed technical inspection to system-level management assessment.

---

[5] The union complaint was finally not investigated by the Ombudsman, who accepted the study commissioned by SodM and described in this paper as a sufficient response to it. The union official responsible for the complaint was interviewed in this study and involved in the feedback process afterwards.

1.2. The industry and the inspectorate

The oil and gas industry in the Netherlands in the early 1990s consisted of 15 operating companies (as defined under the Mining Law), using the services of some 800 contractors and sub-contractors. Some 9000 people were employed in the sector, with about 950 on-shore locations and 100 offshore (of which 40 were manned) in the Dutch sector of the North Sea. The inspectorate dealt principally with and through the 15 operators. It had a staff of 35 (excluding administrative staff). The staff was organised into sections based largely on technical disciplines (see organigram in figure 1).

Figure 1: Organigram SodM 1990

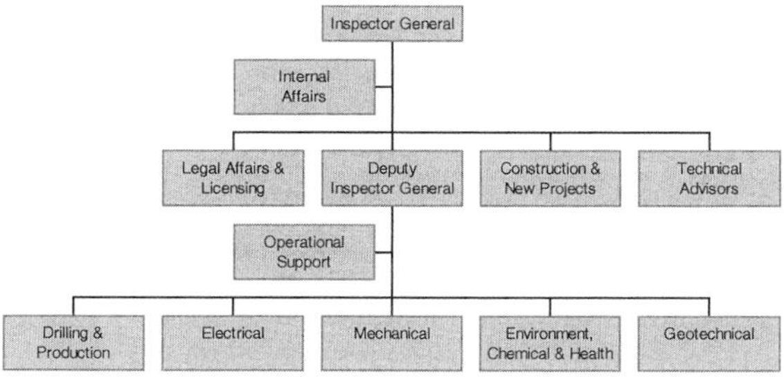

Inspector General
Internal Affairs
Legal Affairs & Licensing
Deputy Inspector General
Construction & New Projects
Technical Advisors
Operational Support
Drilling & Production
Electrical
Mechanical
Environment, Chemical & Health
Geotechnical

In 2000 the sector has declined slightly in size. A couple of operators have withdrawn from the North Sea, leaving 13 active. SodM no longer registers the number of active contractors. The recently very low oil price has meant that there has been less new activity, as the remaining unexploited small fields have been uneconomic to develop.

The inspectorate still has a staff of 35 professionals, plus (temporarily) a number of part-time advisers. The organisation was in the process of being modified during this second study (see below), to create sections paralleling the primary processes of the industry (see proposed organigram in figure 2).

Figure 2: Proposed Organigram SodM 2000

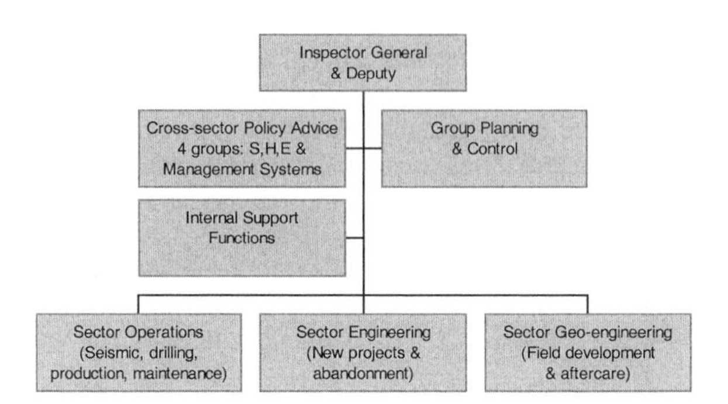

2. THE STUDIES

2.1. Methods, data collection and development of the studies

2.1.1. Initial study 1991 The studies in 1991 consisted of two phases (Hale et al 1992). After initial briefing and study of documents, phase 1 concentrated on SodM itself. The study had been requested by the Inspector General and carefully prepared internally. A supervisory group consisting of him and 8 staff members drawn from all sections and levels was set up to act as sounding board and to ensure co-operation. Interviews were conducted with 9 inspectors and researchers accompanied staff on all their main activities, in order to understand what their goals were, how they perceived the work and carried it out, and what they felt should be the ways of working in the future. The three researchers spent some 15 days with SodM staff in this process. The process was conducted on a very open basis and there was no indication of hidden agendas in the choice of staff to be interviewed, or in the conduct of the interviews.

After a presentation of the results of phase 1 to the supervising group from SodM, and the incorporation of their comments, phase 2 took this picture of current and future working methods to the other actors in the sector. Interviews were held with 6 operating companies and their industry committee (NOGEPA[6]), with 9 contractors, 2 certification agencies, 3 pressure groups and 4 central and local government agencies with tasks overlapping those of SodM. The researchers chose the organisations and the people to be interviewed in them, after discussion with SodM, so as to achieve a representative sample of viewpoints. Interviews were conducted with a guarantee of anonymity, though, in the small world of oil and gas, it is not difficult for insiders to

[6] Netherlands Oil & Gas Exploration and Production Association

guess who has said what. However, this did not appear to hinder anyone in expressing his or her views. Interviews were fed back to the interviewees for checking and correction.
The report of this phase was presented to all of the staff of SodM for discussion and the results were incorporated into a final report with recommendations for changes in the regulatory philosophy and methods and suggestions for changing the structure of the inspectorate to match the new tasks.

2.1.2. Developments 1991-2000 One of the main recommendations was to shift from a bottom-up, hardware-based inspection of the operations to a top-down, system-wide assessment through the SHE (safety, health and environment) management system. This was in line with the decisions already made on the direct basis of the Cullen report. It was clear that this would be a major change for the staff and that they were not equipped to carry it out immediately. It was therefore agreed to develop a training course of 15 days spread over 4 months for all staff. The course was run twice, with half the staff attending each time. The course ran in three blocks of a week, with a month in between in which inspectors applied the lessons learned back in their daily work. After the last block of 4 days the inspectors were asked, in pairs, to assess the recently submitted documents on the safety management system of an operating company and report their findings to their colleagues and to a representative of the operator in a final report-back day. These documents were the first results of the change in regulatory philosophy. A new regulation had required these safety management system descriptions from all operators for the first time in 1992. The study was therefore a learning experience for both companies and SodM. The intensive contacts with the inspectors in this training programme provided rich insights into the problems of the changeover in policy and working methods.

SodM crystallised its decision making about its new role in a document entitled "The New Course" (SodM, 1994). The essence of the new approach was to introduce the principles of system level inspection at the level of safety management systems and the use of risk analysis and safety cases for approval of new developments and for guiding inspection. The progress of the reorganisation was reviewed internally in 1995 and again in 1997. Because of dissatisfaction with the progress of the organisational changes to match the new regulatory principles, SodM commissioned an organisational consultant to supervise a new reorganisation starting in 1999 (SodM 1999, 2000).

2.1.3. The second study: 2000 The data about the situation in 1999/2000 and the way in which the changes in the intervening years had happened is drawn from the internal reports of SodM and the documents prepared to accompany the most recent reorganisation. This was supplemented by interviews with the Inspector General and two of his staff and by 9 interviews with other parties in the field. These were held with NOGEPA and with representatives of 5 operating companies and 3 contractors. The sample was smaller than in 1991, in line with the more limited objectives and funding. The objective was simply to check on how SodM had modified its approach and what this had meant for the organisations in the field in the way of new tasks, freedoms or constraints in organising their risk control. As far as possible the companies and individuals were chosen from the list interviewed in 1991, so that they could directly compare the situation in 1991 and 2000. The reports of the interviews were checked with the respondents. The preliminary conclusions were then discussed with the Inspector General and his deputy to clarify a number of issues.

In the next section a pen portrait will be given of the method of working of SodM in 1991. The paper then presents a framework for considering the role of the inspectorate, expanded from the one used in the initial study. The findings and recommendations of the initial study and the subsequent events and problems with implementation are discussed within that framework. Finally the paper considers some issues relating to the structure and manning of an inspectorate to match its new role assessing SHE management systems.

2.2. Pen-portrait of SodM in 1991

The goals of SodM have not changed over the years. What have changed are the methods used and the emphasis between the different tasks. Its activities can be summarised under four headings, three of which match the steps in figure 2 of Hopkins & Hale's introductory chapter to this book (see in brackets below).

- Supervisory and inspection activities to assess and promote compliance with the law (steps 1, 3 and 5)
- Policy development to assist the sector to develop and implement ways to operate safely and efficiently (step 2)
- Enforcement and prosecution where necessary (step 4).
- A source of information to other regulators and interested parties on mining activities (including a societal "memory" function in relation to abandoned mines)

The fourth activity is rather specific to the mining industry, particularly in its concern with monitoring subsidence from old mines and maintaining records to assist redevelopment work in old mining areas. SodM also has a rather more pronounced role in relation to the cost-effectiveness of the oil and gas industry than inspectorates do in other industries. Its place in the Ministry of Economic Affairs, which licenses and approves developments of new fields and collects the State's revenue from the extraction process, was questioned in the aftermath of the Cullen enquiry into Piper Alpha (Department of Energy 1990) and was an explicit issue in the 1991 study. The question raised was whether such a close association might compromise the independence of SodM. The 1991 study found no evidence of such a problem and saw no reason to recommend shifting the inspectorate to another ministry (as Cullen had done in the UK). In the 2000 study there was some evidence that a related issue was raising its head. Operators expressed concern that the increased emphasis in the objectives of SodM on the monitoring of "appropriate extraction" (an objective linked to sustainability of the industry) was leading the inspectorate to enquire more deeply into the financial basis for decisions about field development, exploitation and abandonment. Companies did not appreciate this delving by "government" into their economics. This aspect of the study is not further developed in this chapter, but a similar discussion in the Norwegian sector is more extensively handled in Hovden's chapter in this book.

The primary activity of SodM at the start of the 1991 study was on-site inspection of locations in order to check if the law was being followed. The Mine Law was a quite generally worded document, well in advance of its time, since it was phrased mainly in terms of goals and general requirements. The words "safe", "appropriate" and "satisfactory" appear over 700 times in its articles as objectives. However, some 70 of these had been the subject of regulations issued by the Chief Inspector as detailed interpretations. The resulting regulations were very detailed and oriented towards hardware and specified detailed procedures and competence. The Inspectorate carried out some 2500 inspections a year, 75% of them routine visits with no specific purpose other than the general compliance check. Inspections occupied some 75% of the time of the field staff. These inspections gave rise in 1989 to 300 improvement actions in letters to the companies, 25 entries into the "mine book" (serious breaches of regulations), but no prosecutions. This work was done by the staff in the inspection sections and was largely conducted individually by inspectors for their own discipline area. A few group inspections were carried out when several inspectors visited a site together, but this was usually tantamount to three or four separate inspections in parallel, rather than a group looking together at one aspect or theme. Communication between the discipline areas was largely via the dossiers on the different operators, into which all inspection reports went. These dossiers also contained the vast amount of routine data on work intentions, progress, activities, etc. which flooded into the inspectorate every week to inform them of the detailed plans and work on all sites.

The other activities of the inspectorate included accident investigation (150 a year), assessment of plans for new projects and granting of licences, exemptions and approvals and general advice and support to companies. Again these were largely concentrated at the level of execution of work processes. About 15% of the inspectorate time was spent on these. In addition the inspectorate was responsible for assessing licenses and approvals related to new developments. A separate legal department (see figure 1) dealt with the paper work for these, and a separate construction section assessed the project plans. The senior staff, particularly the chief inspector and his deputy, spent the majority of their time in visits to company headquarters to review annual plans and longer-term intentions and policy, in national and international committees and meetings with industry bodies and other regulators. Senior staff in the sections also contributed substantially to standards bodies and other fora for the writing of regulations, rules and standards.

2.3. Theoretical Framework

2.3.1. Matrix of tasks In order to structure the discussions and thinking about the role of the inspectorate a simple framework was used (Hale et al 1992). This distinguished three levels of operation of risk control, which can be mapped onto the levels of intervention in figure 1 of Hopkins and Hale's introductory chapter (the words in brackets make the link):

- Execution level (Direct risk control): the daily activities in the full range of life cycle phases of oil and gas, from exploration and extraction, through production and maintenance, to abandonment. This is the level of direct control of risks and encompasses all of the hardware and human actions to keep risks under control.
- Safety management level (Safety Management): the planning, policies and methods of control to achieve the safe operation of the execution level, adding up to an explicit and coherent SHE management system for achieving risk control. The safety case can be seen as a part of this level concerned with showing that a new (or existing) plant is designed and operated to an acceptable level of risk.
- Goals and policy level (Regulation based on Output measures): the establishment and control of overall goals for safety, health and environment.

Each level was depicted as a problem solving cycle (Hale et al 1991) consisting of the steps:

1. Defining desired criteria (goals, rules, performance)
2. Assessment of performance against these, to recognise deviations
3. Understanding the deviations
4. Finding solutions to them
5. Implementing the solutions
6. Monitoring the solutions against the criteria with feed back to step 2

The emphasis in the study was laid on the steps of producing the rules/criteria/goals for defining desired performance at each level and of assessing the performance against these criteria. These two steps (boxes 2 and 3 of Hopkins & Hale's figure 2) covered the main activities of the inspectorate. Such a framework produces a matrix of possible tasks as in table 1.

The thesis of this chapter is that these tasks must always be carried out at all three levels in order to control the risks in an industry. It is useful and interesting to ask which of the actors in the system should perform each of them. Which should be done by an inspectorate, which by the operators in the sector and which by relevant other parties?

Table 1. Matrix of level of control vs. task in making rules or checking compliance

Level	Make[7] & promulgate rules	Assess conformity with rules
Regulatory goals	**A**. Establish goals for safety (risk levels), health and environment (e.g. MAC levels or other exposure standards)	**B**. Check that output goals are achieved
Safety management & safety case	**C.** Make rules for safety cases & SHE management systems and how they control risks	**D**. Check the structure and functioning of SHE management systems
Direct risk control	**E.** Make detailed rules for the execution level	**F**. Check that execution level rules are carried out on site

This matrix maps onto a hierarchical view of rules and their definition (Hale & Swuste 1998), which defines three levels of increasing specification and tightness of control. Defining goals (A) leaves it entirely free as to how the goal will be achieved. Defining procedures to go through in arriving at your own set of rules (e.g. risk assessment procedures, consultation procedures, task analysis procedures) adds some degree of constraint on what are acceptable rules. The procedural level of rule specification matches the boxes C and D in table 1. A management system is one way of defining the procedures a company must go through in defining its own detailed rules. A safety case regulation lays down the procedure for assessing risks in the design stage and demonstrating that they are controlled. Specifying the rules in detail in regulations (E) removes all freedom of decision from the operating level.

2.3.2. Culture of regulated companies Besides the matrix of table 1, we make use in the discussions in this chapter of the framework proposed by Westrum (1991) for characterising company culture related to safety. It provides a useful classification of the stages of development in a responsible and effective risk control policy and management system in a company. The scale ranges from *pathological* to *generative* companies. *Pathological* companies deny risk and try to duck laws on safety. *Reactive* ones respond to accidents and to enforcement by taking corrective action, but do no risk control of their own volition. *Proactive* companies anticipate risk and take steps to prevent and control it, but still regard it largely as an add-on extra. Both reactive and proactive companies are *calculative* (Westrum 1988) in their approach, weighing costs and benefits of action, but they differ in where they place the balance points and what they allow to weigh in the equation. *Generative* companies have taken the final step to internalise risk control in

[7] In the case of laws and legal regulations the situation is in fact more complex. It is the Ministry and its legal experts who formulate the rules and the parliament or Minister who approve them. The inspectorate has formally only an advisory role. In practice, however, it is the inspectorate which determines the content of the legal rules. This is certainly the perception in the field, and this shorthand will be used in most of this chapter.

all their activities and to make it a pre-requisite, rather than a factor to be traded off. The scale runs in parallel with a number of facets of safety management, notably how open the organisation is to recognise and discuss risk and to accept and learn from incidents and mistakes. We shall return to this scale in discussing, in section 4, how the split of tasks in table 1 is influenced by where the bulk of companies in an industry are on the scale.

3. SodM Characterised: from 1990 to 2000

SodM clearly concentrated its activities upto the 1990s under E and F of table 1. E was carried out either alone, or in collaboration with other ministries or other parties in standards committees. The operating companies were consulted about the detailed rules made by SodM, but the inspectorate had the final say. In terms of time of inspection staff, F dominated the activities. There was some activity under A and B, but it was usually rapidly translated into agreed rules under E to achieve A, and detailed inspections of these rules (F), with B acting as a safety net. An example of an attempt to confine actions to A and B was the establishment of a standard for oil concentrations in water (40ppm) dumped overboard from platforms into the North Sea from drilling and production activities. Companies were left to decide for themselves how to achieve this and no detailed rules were made. This was, however, an exception at the time.

In the 1991 studies the majority of both the inspectors and the operating companies and contractors regarded operation by SodM at the E/F level as inappropriate, wasteful of resources and ad hoc. This view was not shared by the pressure groups and union representatives, who saw the power of the inspectorate in just that attention to detail. The majority, however, advocated transferring the focus of activities from F to D, the assessment of safety case plans and SHE management systems. This was also the recommendation made by the researchers at the end of the study and the decision taken by the inspectorate in its policy document on "the new course" in 1994 (SodM 1994). The operating companies explicitly rejected a corresponding shift of rule making from E to C as undesirable; box C was regarded at the time as the exclusive prerogative of the companies. This presents something of a paradox, since assessment demands a measuring stick. The implication of the companies viewpoint seemed to be that, in the absence of SodM rules under C, either implicit rules would have to be applied by inspectors out of their own heads, or externally produced standards (e.g. from ISO) would have to be adopted.
The companies were also not enthusiastic about any more emphasis on A and B.
The arguments presented under each heading will be discussed in section 4 below.

It is clear that SodM has achieved a quite substantial shift in the proposed direction in the intervening years. The figures for 1998/9 record a drop in routine inspections to about 300/year from 1900/year in 1990, whilst project based inspections only dropped from 600 to 300. The percentage of time spent on inspections dropped from 55% to 10% of total inspectorate staff time. The dramatic increase was in the assessment of company safety cases, safety and health documents, SHE management systems and internal audits and the conduct of audits by SodM staff. This rose from 5% to 25% of the total inspectorate resource time and the work has been pushed down to all levels of inspection staff to replace their on-site hardware inspections. At least another 10% of the staff time now goes to direct internal support for this system level approach. This is over and above the activities of the senior staff, which were always more directed at the system level of operation. In addition, a thorough review of detailed rules in 1992 allowed a reduction from some 70 to around 40. Only two new detailed rules had been promulgated at the technical level and one at the management system level (see 4.1 and 4.3 below) All in all, more than 50% of resource time is now devoted to system-level regulation. This changeover was not without serious problems for the inspectorate and the operating companies and contractors. The next section discusses some of these issues in terms of the matrix in 2.3.1.

4. Deciding Who Does What

The matrix in section 2.3.1 was devised in order to structure discussion about who should do what. The assumption was that good control of risk in any sector requires actions at all of these three levels, i.e. tasks A to F must all be carried out by somebody. It is not possible to do without detailed rules and supervision of them. If nobody else makes them, those carrying out the work will define their own and exercise such care in carrying them out, as they deem necessary. The only discussion possible is therefore about which party should perform which task and notably, whether a government inspectorate should be deeply involved, or should leave that aspect to other parties (Hale & Swuste 1998). In the following sections the arguments for and against a role for the SodM in each aspect are presented. These come from the initial studies and from the experience of SodM and the other sector parties since the reorganisation. The matrix will be discussed from the bottom up (from E/F to A/B) since this is the progression which was proposed in the 1991 study and which SodM has largely followed.

4.1. Making execution rules (E)

4.1.1. Expertise and responsibility The main arguments advanced in 1991 against inspectorate involvement in this step were that it lacks the up-to-date expertise to do so. The other actors in the sector saw SodM as experts in the law, and in the hazards of the technology. They trusted them to adapt generic regulations imposed from elsewhere (e.g. the EU Framework directive) to the peculiarities of the mining industry, but did not see them as experts in the primary technology. The fear was that they constantly lag behind the latest developments in technology and may even put a brake on such developments with detailed execution rules. Regulation at this level by SodM was also felt to remove the responsibility from companies to think for themselves and to propagate a frame of mind in the companies which concerns itself not with whether a measure will work in controlling risk, but whether it will comply with the law. The Mine Law tried to resolve this by using words such as "suitable", "adequate", "well-constructed" and "practicable". However, in the course of time the inspectorate had issued and updated many detailed regulations explaining what these meant. These had often been requested by the companies, who wanted to know what would be regarded as meeting the law. There was still a minority of companies in the 1991 survey who still demanded this and was unhappy with the thought of having to devise their own rules. In Westrum's terms (1991), these are companies still working in the reactive stage of development. In that stage they need something to react to and they resist being pushed into the next stage of maturity as proactive companies. In the area of environmental laws, which, in 1990, were relatively new for offshore, the majority of companies criticised this "vague, objective-setting law" as being too much for them to cope with. This sort of response raises the important question of how rules can be written which, at the same time, allow proactive (or generative) companies to establish their own rules and methods (in boxes C and E in table 1), whilst keeping reactive (and especially pathological) companies under control. Does a shift from rule checking in box F to box D encourage companies to move from reactive to proactive, or does it allow them to slip back into inaction?

In 10 years SodM has reduced the number of their detailed technical rules by over 40%, in discussion with industry. The tenor of opinion is that this has been a successful process in general, but that it has revealed a number of issues and problems.

4.1.2. Industrial fora The experience of the intervening years since 1991 has been that the minority of companies still resist being left to devise their own rules and that SodM still suffers pressure to write detailed rules. However, it has been largely successful in resisting that pressure by directing the demands into the channels of joint rule making by industry bodies and standards organisations. The industry fora, such as NOGEPA for mining companies and the IADC[8] for drilling contractors, fulfil this role to some degree. SodM is prompting them to take over a number of their rules and update and add to them. This process is not a rapid one. Such organisations have to cope with considerable differences of opinion among their members, making it difficult to cut the Gordian knots as fast as SodM could do in the past. Industry bodies in general may have problems being proactive, since they tend to travel at the pace of their slowest member, unless the progressive companies dominate and have sufficient hold over their more reactive fellows. There is also potential conflict between NOGEPA and the contractors' federations, because of different, or even conflicting interests, which may necessitate some intervention from outside to resolve it.

[8] International Association of Drilling Contractors

4.1.3. International Harmonisation Alongside the Dutch industry bodies there are many international fora in which detailed rules and standards are made. An important issue requiring SodM activity in box A has been this issue of international harmonisation. Some aspects of operation must be standardised at a detailed level across all sectors of the North Sea to avoid confusion. This is true of such issues as emergency procedures, alarm signals and helipad lighting, which are all used by companies and services operating across national sector boundaries. Another reason for cross-national harmonisation is to improve controllability, e.g. in training requirements and courses for the safety-training passport, to allow free movement of personnel between jurisdictions. This harmonisation issue has meant that SodM has continued to be involved in detailed discussions on rules in the North Sea Offshore Authorities committee. This seems to be an example of regulators retaining detailed involvement when such negotiation and harmonisation could be left to the operators and certifiers and simply rubber stamped by the authorities. However all jurisdiction authorities need to agree to that delegation to the sector before it can happen. If one does not accept, then the other jurisdictions more or less have to send their regulators to negotiate on an equal footing.

4.1.4. Competitive pressures and "appropriate rules for the sector" A further reason given by SodM in 2000 for some continued involvement in detailed rules was the prevention of competitive advantage. The case was quoted of the maximum number of beds permitted per room in the sleeping accommodation on platforms. A number of years ago (still under the remains of the old detailed regulation regime) operators were informed that this maximum would be reduced from four to two, with a lead time of several years to allow design modifications. Operators had spent considerable sums of money making these and it was necessary for SodM to enforce the rule, because new operators coming into the sector would be at a considerable financial advantage if they did not have to comply with the rule. This example demonstrates the intimate link between safety rules and competitive and commercial pressures. The evidence from this study is that SodM is well aware of this link and prepared to adapt its strategy of risk control to avoid unnecessary costs and provide a level playing field for competition. Indeed, companies interviewed in 1991 felt that this was one of the main advantages of an industry-based inspectorate such as SodM over a more broadly based general inspectorate such as the labour or environment inspectorates. SodM was able to tailor the detailed rules to the realities of the industry, which they knew well from having worked in and for it for so long.

In addition to industry specific and international fora for rule making, other Dutch ministries play a role in making detailed rules. In the 1991 study there was considerable discussion of this issue. It was recommended that SodM should supplement its expertise by drawing on that of the Environment and Social Affairs Ministries in relation to environmental and working conditions legislation respectively. Such a move was welcomed by the industry parties, provided SodM retained its role in translating rules to the specific circumstances of the oil and gas industry. In the ten years since the first study this process has gathered pace. Under European and national pressure there has been a move to harmonise legislation not only across countries, but also across sectors within a country. Large parts of the Dutch environmental and working conditions laws are in the process of being declared applicable to the offshore continental shelf. Although all of the other Ministeries were already parties to rule making before 1991, their role has now become more visible and their voice correspondingly stronger. This has led to more voices in the 2000 interviews calling for a specific translation of these rules for offshore conditions. In that sense the demand for an independent SodM has strengthened. However there is another side to the issue.

With the withdrawal of SodM from such an active involvement in the detailed rule writing, as described in the earlier paragraphs of section 4.1, their power to fulfil this translation role has become diminished. All parties were agreed that this withdrawal had, indeed, taken place to a significant degree. However, rather than leave this level entirely to the other actors, SodM has tried to retain a guiding hand. It participates in many of the bodies mentioned in order to ensure that the rules made do not accept too easily a majority view, which might water down safety levels. In this way it has tried to respond to the concern expressed by the government and pressure groups interviewed in the original survey in 1991. It is also the principal adviser of the Ministry of Economic Affairs in the process of adapting legislation from other sectors for incorporation in the new Mine Law, being prepared at present. From this position it can watch over the appropriateness of the rules for the oil and gas industry. However, there was a clear increase in the level of concern among industry representatives, interviewed in 2000, as to whether it was still in a powerful enough position to fulfil that role at the detailed rule level. If that were proved not to be the case, a number of the interviewees felt that it would largely remove the argument for an independent inspectorate and favour its merger into another ministry. It would appear that the proliferation of rule-making fora at this level, coupled with the lower profile of SodM in these fora, has resulted in a certain loss of faith in the power and focus of the inspectorate.

4.1.5. Freedom of choice for the sector One final reason for delegating detailed execution-level rule making from the inspectorate to the sector was to allow more freedom of choice to companies in how to innovate and to arrange their processes. This should allow the introduction of new methods and technology where appropriate (see v.d. Poel et al 2001 for a more detailed discussion of this aspect of the study. Baram, in his chapter of this book, also discusses the relation between innovation and regulation in detail). Looking back at 5 years experience there appeared to SodM and other sector players to be only a small amount of evidence that this was happening. The cases quoted included:

- The acceptance of underbalanced drilling[9] on the basis of explicit risk analysis, which had been banned under the detailed rules.

[9] Drilling of wells with less than the previously required pressure to prevent blowout. That required pressure had been based on worst case experience.

- More flexibility in inspection schedules for pressure systems, based on risk based arguments, after the withdrawal of the rigid time interval requirement[10].
- Stimulation of the industry to come up with radical new technology or solutions to achieve the 40ppm oil in water standard, and the enormously reduced radiation standard (from 500 to 1 beq/gm) for drilling or production spoil[11].
- The development of production platforms that can be reused in a series of small fields, something previous rules did not permit.

This is not a long list. On the other hand, perhaps 5 years is too soon to make a judgement on this issue. However, the parties interviewed felt that the change in SodM rule making from E to C had not resulted in much additional actual freedom for manoeuvre and for different technological or procedural solutions between companies because of other effective constraints, which are limiting the use of this freedom. There was already much freedom of choice under the old regime, because of the flexibility of interpretation of the inspectorate. There is now much more company benchmarking operating through the industry bodies, such as NOGEPA. There is also the "procedural" regulation to use "best practicable technology", which encourages this process. More important for many interviewees was, however, the belief that the industry in the North Sea was largely mature and not suited to radical innovations, because it is reaching the end of its life span. In particular, safety measures were felt to be more or less fully worked out, so no innovation was necessary. In any case, the North Sea industry in 2000 would appear to contrast strongly with the biotechnology industry described by Baram in his chapter, where innovation has the upper hand in a climate of relatively weak legislation. Such "wild west" days in the oil and gas industry were several decades ago (see Hovden's chapter).

4.2. Enforcing detailed execution rules (F)

Government could retain a role in enforcing detailed rules, even if it were not to be involved in writing them. In a sense, this is what SodM now does in auditing the SHE management systems of companies. It looks to see if the detailed rules devised by the companies are being applied by them. The contrast with the former practice is that the SHE management system now acts as framework for the inspection. One of the concerns of some of the less proactive companies in the initial survey was that SodM would use the best companies as their reference standard when assessing them, rather than using an acceptable minimum compliance with the law as the measuring stick. With the development of detailed rules by NOGEPA and other standards bodies, there is now an

[10] This is an example of a rule which has not only been relaxed, but also handed over to the independent certification bodies to check. SodM accepts this step, but objects to the subsequent proposal that the large companies should be allowed to self-certify their inspection processes, using their own inspection services, subject to accreditation of these.
[11] However, the proposed technology of re-injection of spoils in the old wells, found acceptable by both industry and SodM, was rejected by the Environment Ministry as against one of their primary sustainability goals of retaining the recycling option for all radioactive waste.

independent set of rules to be enforced. SodM in 2000 would appear to see this to be their role. They plan to use the agreed industry rules as the default test of the rules developed by individual companies; the company's own rules need to be at least as good. Some operators regard this as something of a threat to the essentially voluntary nature of the rules and as proof that SodM has not sufficiently withdrawn from all concern with level E/F. It is inevitable, however, that enforcement, by whatever body, has to have some agreed set of rules as its yardstick.

The objections to the enforcement practice of SodM, as it was in 1991, were that it was very arbitrary. Inspectors were not systematic in checking the rules, but used a "gestalt" of what they thought to be good practice. Walking round the site led them to detect deviations from this, which they then brought up with the operator or contractor. What deviations there happened to be on that day were felt to dominate the inspection. In addition companies accused the inspectors of having idiosyncratic hobbyhorses which they pursued. Only rarely did they seek the deeper causes underlying the deviations and penetrate through into the management system. Their approach was also reactive in Westrum's classification. The other side of the coin to this arbitrariness was the belief that SodM was flexible and had an eye and ear for practice and the necessity to continue in production. Inspectors were therefore not too rigid in their interpretation of the detailed requirements, because of their knowledge of the industry. Operators expressed fears that this flexibility, based on long experience, would be lost if the inspectorate amalgamated with the general labour inspectorate and lost its specialised personnel and nature (see 4.1.4. above).

The loudest voices raised in 1991 for retaining a government role in task F were those speaking for the unions and pressure groups. They were concerned that, otherwise, only the companies would make the standards and their compliance would not be checked rigorously. They feared that this would not give such great protection to their constituents. Their point of view was supported by some SodM inspectors who feared that they would be less able to pin down and achieve change in, or even prosecute poorer companies, if there were no formal execution rules in the government regulations. All actors also recognised the advantage of an external body occasionally checking up on detailed practice, to uncover things that routine has blinded the company to. The more reactive companies also welcomed the inspectors visits as a way of keeping the workforce on their toes, a task they did not appear to welcome doing themselves. The general consensus, as solution to this need to retain a regulatory external check on detailed rules, was to steer this task in the direction of independent third-party certification, based on (international) standards. This has, to an extent, been done, but SodM in 2000 still sees a continuing role for itself in this task (see above).

Another argument advanced for a continuing involvement of inspectors in E and F activities was the need for inspectors to retain contact with and detailed knowledge of the technology and the reality of the industry. This is comparable to the "situation awareness" needed by traffic controllers to manage the complexities of air or vessel traffic flows: without it, it is difficult and even dangerous for an inspectorate to try to steer or control the oil and gas industry. The retention of a minimum of routine inspections and field visits by SodM is an attempt to retain this contact, as is their requirement for applicants for inspectorate posts to have 10-15 years experience in the industry. However, the interviewees in 2000 exhibited more concern than those in 1991 about the success of SodM in retaining this contact and sense of reality. Inspectors were felt not to have the broad experience found previously; there was a concern that too many came from a narrow range of large oil and gas companies, rather than the broad spectrum active in the field.

4.3. Writing rules for SHE management systems (SHEMS) (C)

In the 1991 survey there was a general rejection of the idea that SodM, or indeed anybody else, should write rules at any level of detail on how management should manage (box C). This was seen by operators to be their inalienable prerogative, the freedom to manage. Well-known arguments were advanced about the variety in acceptable ways to manage and cultural differences in management styles and structures.

The inspectors themselves did not want this power to write or enforce detailed rules about management structure, because they recognised that they did not have the expertise to exercise it. However, SodM then found itself in a double bind in relation to the task it did wish to carry out, namely the checking of the management systems (box D). Without agreed definitions of what to look for, it is impossible to check what is there. It asked initially for descriptions of the SHEMS, without giving any indication of what, in detail, it wanted to see. When pressed for more information, it issued a guideline for notification. It was then criticised for being too prescriptive. In the intervening years the debate has continued, aggravated by the continuing absence of any accepted national or international standard for at least the S and H parts of the SHEMS. Without an agreed standard against which assessment can take place, it is inevitable that the practice of the regulator in assessment will define the rules. The production and circulation to the operators of a SodM document (SodM 1996) detailing lessons learned from the assessment of SHEMS in the first round of reviewing company management systems gave rise to (for SodM) unexpected furore in the sector. It was seen as a continuing attempt to impose rules, and, perhaps more importantly, as damaging to the image of the sector, because it concentrated on the negative aspects of the experience. It would appear to be the companies coming from the American culture who are the most resistant to anything approaching rules of management. They are concerned about the implications for liability (Baram 1997).

In 1995 SodM decided to formalise the practice which it had been developing up to then. It issued a rule under the Mine Law (Ministry of Economic Affairs 1995) defining how it would approach assessment of safety and health management systems. It declared a number of ISO standards as relevant for designing and auditing the management system. These were quality management standards (NEN-ISO 9001, 9004, 10011, all in the 1994 edition). A number of companies objected to this quality management approach as being not entirely appropriate for health and safety management. There has been a lively debate in the literature about a number of possible objections to such a use (see e.g. Zwetsloot 1994, NEHEM 1994, BSI 1996, Cottam 1999), on grounds such as the difference between product and process control, the different involvement of employees and different social and regulatory approaches. A number of respondents questioned why the ISO 14001 Environmental Management Standard had not been declared relevant and why equivalent standards (such as OHSAS18001) were not permitted. This lack of flexibility in the regulation would appear to have been imposed by the legal advisors of the Ministry in drafting the rule. SodM, in letters in answer to questions from some operators, has indicated its preparedness to accept such alternative standards, since they are based on the same management principles. However, the message seems not to have penetrated the sector as a whole. The main objection of the oil and gas companies has, however, been centred on the interpretation of the requirement under the standard to define their business/work processes and the risks related to them. This point is developed further in section 4.4 below.

The imposition of the NEN-ISO standards and the requirement to define work processes appears to have been interpreted by the industry quite widely as regulation at too great a level of detail. In this it is seen to be an approach reminiscent of the old detailed technical rules, but now applied to management. It may be that some of the objections are simply a general resistance to change: "if it was not necessary before, why is such detail necessary now?" The concern about detailed management rules may also reflect a question dealt with in more detail in the next section. This is the fear that companies have that inspectors do not have sufficient management knowledge to be able to interpret and apply the detailed management rules as flexibly as they used to apply the detailed technical rules.

It is interesting that this discussion on management rules should be so central. The Mine Law had always had many articles formulated in generic ways requiring suitable hardware, competent persons and adequate procedures. These imply management systems to produce and achieve them. However, SodM had always translated these generic requirements into specifications of the results of the management processes, not rules about those processes themselves. It is the step from defining (and assessing) the product to defining the process which was seen as revolutionary and an unacceptable task for the government regulator.

4.4. Assessing the structure and functioning of SHE management systems (D)

This section covers two aspects of the proposed new course for SodM, the assessment of design safety cases based on risk analysis and the assessment of safety management systems. The two aspects were linked in the "safety case" notification requirements of SodM. Operating companies have to submit both as concepts for checking and comment by SodM.

4.4.1. Safety cases The first aspect, the technical design case and risk analysis has been the least controversial. It had already been an element in the old task of SodM and the sector valued the practical input of the SodM assessors, based on long experience and wide knowledge of industry best practice. There was not the concern about SodM creating their own rules for this aspect of the assessment, in contrast to the attitudes to management rules. Nor was there any proposal to delegate assessment to certifying bodies. This reflects a different level of trust in the technical as opposed to management competence of SodM. A new element that did concern the sector was that the first safety case information had to be submitted in the early concept design stage. There was a certain lack of trust of SodM with such commercially sensitive ideas, which could determine competitive edge. The general consensus was that the first safety cases were much too detailed and that both sides had to go through a learning process to arrive at a suitable compromise in the level of detail. However, most people on both sides now regard the safety case regime as a success, which has made the industry think much more clearly in terms of risk, rather than of deterministic rules, as they used to do. Their risk control measures are therefore more transparent. This can be seen as a shift from rules at level E up to at least C, if not A.

SodM has found itself in a number of cases defending absolute rules, whilst the industry argues for risk based ones. The controversy over the presence of standby boats for rescue at all rigs is a case in point, where the operators argue that risk analysis shows it not always to be necessary, whilst the inspectorate wish to insist on it as a pre-requisite. Hovden's chapter discusses this issue of risk-based vs. pre-requisite rules in more detail, since it seems to have been a much more explicit debate in the Norwegian sector.

The concern about access to sensitive commercial information, previously seen in the safety case discussions, has resurfaced recently in the debate about safety management system assessment. Companies have been reluctant to divulge financial information about the grounds for decisions on investment related to safety and to the exploitation of marginal fields.

4.4.2. Safety management systems The interaction between assessing the SHEMS and writing rules for it has been indicated above. In 1991 SodM, the research team and the other actors in the sector all recognised that there was too little expertise and experience in SodM to carry out this management assessment task. However, all agreed that it was the task that should in future be central to the inspectorate. Some consideration was given to delegating this process to certifying bodies. However, there were, and still are, only standards for quality and environmental management systems and not for safety and health.
Questions were also raised, particularly within SodM about the independence of the certification bodies, given the fact that their services are paid for by the companies and that there is a competitive market for certification. They felt that management systems was an area open to such

variation in interpretation that it was not advisable to leave it to such a free market. Gundlach's chapter in this book provides some further discussion on this issue.

A point of discussion was whether SodM should conduct its own audits of the SHEMS, or should rely on the documentation and the company's own internal audit. The lack of confidence in their own expertise in this area led SodM to prefer the latter course. Companies feared that immediate access to these internal documents would compromise their honesty and usefulness to the company itself. They were only prepared to grant it if sufficient time was allowed to put right shortcomings found. With this compromise the system appears to have been accepted and only a small amount of time of inspectors is spent on independent audits. This raises, however, the question as to how penetrating and independent these assessments of the SHEMS really can be. There is a major difference between the insights to be gained from accompanying company auditors and from planning and conducting a fully independent audit (see e.g. Hale et al 2000)

The lack of expertise of SodM, noted in 1991, to carry out the SHEMS assessments implied the need for new recruitment and training of all staff. New staff was indeed appointed and a new section was set up to coordinate the assessments and advise existing inspection staff on it. The 15-day training programme was set up to provide all staff with the basics of the assessment. This proved to be a much more fundamental and difficult process of individual and organisational change than at first thought. It was already clear in the training course that some inspectors took to the system-wide view and the reversal of approach from bottom-up to top-down inspection like ducks to water. Others found it far too great a culture shock, not compatible with their technical interests and competence, a threat to their authority, or a waste of their many years of accumulated experience with the old approach. The experience of the intervening years has shown that a number of inspectors did not succeed in making the change of approach and either left, retired early, or were moved into functions which could still use their old talents and approach. Only with the coming of new blood has the reorganisation really taken root in this respect.

4.4.3. Organisational change in SodM The organisational changes were also made difficult because of a lack of clarity about how to recast the structure of the inspectorate. It had been organised on a technical discipline basis (see figure 1, section 1.2), each discipline inspecting more or less independently of the others. The researchers proposed in 1991 a radical change in this structure, with a multidisciplinary general inspectorate using the assessment of the management system as its starting point and calling in specialist, second-line, technical experts for problems outside their competence. This was too radical for SodM. A compromise was introduced, with a staff section advising and coaching on safety management and a minor reorganisation of the discipline sections, with an amalgamation of mechanical and electrical engineers into an operations section. The staff section was supposed to prepare the work of the technical sections, by analysing the SHEMS documents and proposing areas for study. Group inspections with several technical sectors involved were supposed to provide some coordination of approach. The safety management section was then supposed to coordinate the information collected into a coherent assessment. This compromise did not sufficiently recognise the radical nature of the reorientation of the inspection work. Inspectors, the majority strong-minded individualists, did not adapt their approach sufficiently and did not take to the idea of group inspections. The internal reviews in 1995 and 1997 showed that the system worked on paper, but not in practice. In 1999 SodM (SodM 1999, 2000) recognised that the compromise structure was too much determined by historical precedent and not adapted to the primary tasks of either the sector or the inspectorate. A new structure has been proposed and implemented in 2000, which takes a further step towards the more radical 1991 proposal (see figure 2, section 2.1). It maps SodM onto the primary processes of the sector. It allocates field development and aftercare to a Geo-engineering section, design, construction and abandonment to an Engineering section, and seismic, drilling, production and maintenance to an Operations section. Each section is responsible for looking not only at the technical aspects, but the environmental, health and management aspects. Four project teams drawn from the new sections, one each for safety, health and environment and one for management systems, will have the task to coordinate on these system-wide aspects. A small advisory staff section supports them in the management aspects and in keeping up-to-date with external developments in all the system-wide tasks. The hope is that each section can thus look at the management system for a coherent part of the life cycle. With the introduction of this structure and of new staff with no "historical baggage" wedded to technical inspections, it is hoped that assessment of management and a system-wide perspective will become central themes for each section. Only time will tell if the inspectors can adopt this approach fully, or whether the technical disciplines will still dominate over the need to look at the technology through the spectacles of the management system.

4.4.3. Detailed management requirements Meanwhile, the assessments being made of the management systems are being guided by the 1995 rules (see section 4.3. above). The most controversial element in these is the interpretation given by SodM to the requirement to identify the risks in terms of the primary business processes per life cycle phase. This is designed to act as the basis for defining the control measures and management control systems needed (see also Hale 2001 for a theoretical underpinning of this approach). It is a central requirement of all ISO management standards, which has been reaffirmed and strengthened in the recent ISO9000:2000 standard. Many companies claimed that the degree of detail asked for by SodM was excessively bureaucratic. It has required them to conduct many man-months of work making far more detailed analyses of processes, which they felt were already well controlled. The response of SodM is that they have not *imposed* such a detailed level of analysis. They have simply, for their own use as a check instrument (under D), made a very detailed process analysis to aid inspectors in their dialogue with companies. They indicate that they have accepted much less detailed analyses when companies have submitted them and argued their case. This debate, and the example of the ISO standards quoted in 4.3, illustrate an interesting phenomenon in safety regulation. The perceptions of what the rules are (C) and how they will be interpreted (D) are quite diverse across companies in a system where the rules are less sharply defined than before. There seems to be a tendency then for many companies to hang on the lips of the inspectors and to interpret their suggestions as requirements. They do not challenge what they see to be the inspectorate's wishes, preferring the line of "appeasement" and the quiet life. In this way the word of the inspector becomes law. This is a typical response of reactive companies in Westrum's classification, which have not yet learned to formulate and defend their own risk control approach.

There are some signs that the issue is resolving itself. Companies that have conducted the business process analysis in detail have found more value in it than they expected. It may be that the same process of learning about an appropriate level of detail is taking place in relation to safety management systems as took place with safety cases (see also Visser 1998 for a further discussion of detailed risk analysis and management systems).

4.4.4. Contractors Another issue which falls under this heading of assessment of management systems was of great concern in the 1991 survey and is still so. This is the control of contractors. Under the Mine Law the whole responsibility for risk control is placed on the mining organisation (the 15 operators in 1991). It is up to them to assess and control the work of all contractors they hire in. This had meant that SodM had not had intensive contact with contractors in making rules, and would pursue the operators if it found contractors in breach of regulations during an inspection. The contractors resented the lack of contact in the rule making and considered that they should have an equal voice with the operators and their industry body (NOGEPA), certainly where it concerned execution rules they would be expected to carry out. They also considered that the lack of harmonisation in both these and the management systems of the operating companies put them to great cost and created staff confusion as they moved from site to site. SodM accepted this criticism and has made considerable attempts to consult with contractors since then.

In 1991 there was a complex system of A- and B-contractors in operations. The former were considered large and mature enough to manage their own activity, and hence to have their own management system assessed. The latter were considered to be too small to have a full SHEMS and hence should be considered as falling under the management system of the operator or main contractor. Designation as A or B was not well defined or controlled. Operators could even decide for themselves, in some circumstances, that a contractor should be designated an A-contractor and thereby avoid any management responsibility for it. The initial study proposed that designation should be tightened up, that the status should be defined per contract and not per contractor, and that the default status should be B unless the operator could demonstrate that the contractor was capable of self-management. In other words, the proposal was to make assessment of the SHEMS of contractors a task for operating companies, which they could delegate to third party certification using the VCA certificate (SSVV 1997). In this way, the contractors would be made more accountable, and could be pursued if they failed to implement their own SHEMS. The researchers and SodM felt that supervision purely top-down via the operating companies would leave too many links in the chain before arriving at the sub-sub-sub-contractor. It was too much to expect all companies, especially the more reactive ones, to control this chain actively. The proposal was to make both contractor and the commissioning organisation responsible and pursue both for their respective responsibilities.

This is largely the line that SodM has tried to follow. However, it has run into a problem with the public prosecutor who has stuck to the old interpretation of the Mine Law and has refused to pursue contractors for breaches of their own rules, prosecuting the operating company instead. The operating companies have largely adopted the VCA certification of all contractors and so have taken the task of box D onto themselves, rather than leaving it to SodM. This of course plays into the hands of the prosecuting officers in the above cases.

4.5. Establishing and checking goals for risk control (A/B)

Goals indicate what has to be achieved but not how. A goal can be *that* the risks are managed, leaving the definition of *how* to the next level of rule making (box C). The problems of withdrawing government regulatory control to this level are largely in defining goals in a measurable way, which does not reduce them to the procedural/management or execution rules of boxes C and E (Hale & Swuste 1998). This is possible in some cases for health and environmental standards, expressed as exposures or maximum allowable concentrations. Examples quoted above, where SodM has adopted this approach with some success are the standards for oil in water and for low activity radiation. It has not been popular with companies, just because it has forced major rethinking of technology to achieve them. This has as much to do, perhaps, with the tightness of the standards imposed as with the principle of using such goals. In 1991 the majority of operating companies interviewed did attack the oil in water standard as too vague and ambitious, but they seem to have responded effectively to it. A complicating issue with this standard is that it is interpreted differently by the authorities in different jurisdictions, with the UK sector adopting an interpretation (at the level of boxes C and D) that allows more contamination than in the Dutch sector.

It is more difficult for safety risks, although risk contours are used for major hazards. In the Norwegian sector there has been a much more explicit debate about the use of company defined risk goals. This is discussed in Kjellén & Sklet (1995), Kjellén (1995) and in Hovden's chapter in this book.

An additional objection from the inspectors in the 1991 survey to restriction of SodM's task to this level was that it would remove them from all contact with the sector and with its real problems. It would make them pure administrators, measuring and punishing. None of them, not surprisingly given their background and the type of pleasure they found in their work, wanted this.

It provides an interesting coda to this section to note that a different Dutch Ministry, Transport and Waterways, has a proposal under consideration to make just such a split in relation to the regulation of safety and environment at Schiphol airport. In the debate around the recently published policy document on the future of the airport (RLD 2000) it has suggested that the inspectorate should concern itself purely with assessing whether the airport and its operators stay within the defined noise, pollution and external safety standards. All of these would be defined in terms of goals giving a permitted risk volume (box A), with attached measurement methods, which could be defined and assessed as "procedural rules" (box C/D). Operators would be free to use this risk volume to the full. It was seen as a way of forcing technical and procedural innovation (box E) to reduce risk at source. Of course, for the residents exposed to the risk, noise or pollution, this innovation would not result in lower risk, since the airport would fill the risk volume released by increasing the number of flights. The actors concerned would also be required to have a safety management system, which would be independently audited and approved. The contents of the management system would be assessed in box D, but the presence of the certificate would be assessable in box A. All discussion and assessment of *how* the management systems and the technical, procedural and operational rules should be formulated and assessed would be delegated to other bodies. A candidate for the integrated management system assessor is the Safety Advisory Committee for Schiphol (see Hale 2001), with independent certifiers assessing the SHEMS of the individual actors. Boxes E and F in table 1 would be left to the companies and third party certifiers, under the general coordination of the integrated safety management committee. This would be a radical regulatory departure and much debate will have to take place before it is clear whether such a step has either the necessary political support or rational basis for success. No such proposal has been made for the oil and gas sector. Indeed the developments described by Hovden in his chapter, which are happening in Norway, would suggest that the trend in the North Sea may be away from global risk acceptance criteria and back to a safety net of pre-requisite rules at operational and technical level.

5. Conclusion

This chapter has reviewed ten years of study, debate and change in the role and structure of the Dutch mining (oil and gas) inspectorate. It has attempted to show, by means of the case study, the range of issues which arise in deciding on the type and level of regulation of one industry. The concentration has been on the role of the government inspectorate in this, with other options being the actors in the sector themselves, and other regulators, such as certification bodies. The theme has been that, if we want effective risk control, we do not have the choice *whether* to have rules at the three levels depicted, only *who* should make them and enforce them. This is a question of where we want to place the debate between the regulator and the regulated, and what we leave to the discretion of the regulated. This chapter has concentrated on boxes 2 and 3 of Hopkins & Hale's figure 2 and not touched at all on boxes 4 and 5.

Implicit in the choice is the question of trust and clarity. The less the level of trust, and the more the process of regulation is adversarial, the more that regulators are drawn to defining or adopting detailed execution rules in laws or standards. Westrum's (1991) pathological and reactive companies can be nailed by them. Lawyers and inspectors without a high level of training can understand them and judges can sentence on them. If the debate is to take place at the level of management systems and generic "procedural" rules, it demands a degree of maturity of, and trust between the partners, which is at least at the lower boundary of Westrum's "proactive" management system category. . A debate at this level also demands a high level of expertise in the regulatory staff. They must be able to interrogate the proposals for rules and the structure and functioning of the management system made by the regulated and assess them on their quality, without recourse to a black and white template of acceptability. In particular we have shown that there needs to be a satisfactory compromise reached about the level of detail at which the procedural rules in a safety case or safety management system will be defined and interrogated. An agreed international standard and protocol for auditing specific to health and safety management systems would simplify this debate.

For an industry such as oil and gas, which has a high profile on risk, because of disasters such as Piper Alpha, and which has a relatively restricted number of actors in the Netherlands, it is feasible to aim for this degree of trust and confidence. It is much more questionable whether it can be applied successfully to the much more diffuse general Working Conditions law, which must be applied in companies ranging across the full spectrum of Westrum's company types from pathological to generative. The contacts between regulator and regulated necessary to negotiate the rules, in the way in which SodM can do it in a small industry, are very difficult to arrange. The Dutch experiment in general industry of delegating a large part of this process to third party certification and the intervention of compulsory working conditions services as (surrogate) regulators is an attempt to do this which has still to be evaluated for its success (see also Gundlach's chapter in this book).

Our study has shown that SodM has been largely successful in shifting its main focus from detailed rules to the procedural level of rules and assessment. The road has not been easy and is not complete. The effects of the substantial withdrawal from operational rule-making have been largely, but not totally positive. The greater number of bodies involved in that level and the lower profile of SodM have tended to make it less visible. This has raised questions in the minds of some of the sector players, whether it still has the power to play a central role in adapting legislation to the peculiar needs of the industry. This has been a major plank in the good relations between regulator and regulated in the past. Issues that are still not resolved are:

- Exactly how an inspectorate can and should function in boxes D and F when it has only a limited say in the way the rules in boxes C and E are written.
- The appropriate level of detail at which procedural rules should be written and interpreted for defining what is a good safety case and a good safety management system.

However, the assessment of the results of the changes over the past ten years also has many positive aspects Companies have become more risk aware, safety cases and safety management systems are positive developments and there is more understanding of how and why risk control and management measures work.

6. REFERENCES

Arbeidsomstandighedenwet (Working Environment Law) 1980. Staatsuitgeverij. s'Gravenhage. 1980

Baram M. (1997). Shame, blame & liability: why safety management suffers organisational learning abilities. In Hale A.R., Wilpert B. & Freitag M. (Eds.) *After the event: from accident to organisational learning.* Pergamon. London. Pp161-178.

British Standards Institute (1996). BS 8800: *Guide to health and safety management systems.* London. British Standards Institution.

Cottam M.P. (1999). Certification of occupational health & safety management systems. In Schueller G.I. & Kafka P (Eds.) *Safety & Reliability.* Balkema. Rotterdam.

Department of Energy. (1990). The public enquiry into the Piper Alpha disaster. (*Cullen report*). London. HMSO.

Hale A.R. (2001). Regulating safety at and around airports: the case of the integrated safety management system for Schiphol airport. *Safety Science* (**37**) 2-3 pp. 127-149

Hale A.R. (in press). Management of Industrial Safety. Chapter V1.0.3. of the *Encyclopaedia of Life Support Systems.* UNESCO. Geneva.

Hale A.R., Goossens L.H.J. & Oortman Gerlings P. (1991). *Safety management systems: a model and some applications.* Paper to the 9th NeTWork workshop on Safety Policy. Bad Homburg. May.

Hale, A.R., Goossens L.H.J. en Timmerhuis V.C.M.. (1992). Staatstoezicht op de mijnen - een verkenning van rol en toekomst (State Supervision of Mines - an assessment of their role and future). *Report to the Ministry of Economic Affairs.* 2093-2098

Hale A.R., Guldenmund F., Goossens L.H.J. & Bellamy L.J. (2000). Modelling of major hazard management systems as a basis for developing and evaluating tailored audits. In *Proceedings of the 1st International Conference on Occupational Risk Prevention.* Mondelo P.M., Mattila M. & Karwowski W. (Eds). ISBN: 84-699-1242-9

Hale A.R. & Swuste S. (1998). Safety rules: procedural freedom or action constraint? *Safety Science.* **29** (3) 163-178

ISO 1995. *Environmental Management Systems - Specifications with Guidance for Use*. (ISO 14001). Geneva. International Standards Organisation

ISO 9000:2000. Geneva. International Standards Organisation.

Kjellén U. (1995). Integrating analyses of the risk of occupational accidents into the design process -- Part II: Method for prediction of the LTI-rate. *Safety Science* **19** (1) 3-18

Kjellén U.& Sklet S. (1995). Integrating analyses of the risk of occupational accidents into the design process Part I: A review of types of acceptance criteria and risk analysis methods. *Safety Science* **18** (3) 215-227.

Ministry of Economic Affairs. (1995). *Nadere regelen Mijnreglement continentaal plat: veiligheids- en gezondheidszorgsysteem.* (Further rules under the Mine regulations for the continental shelf: safety and health management systems). Den Haag. Ministry of Economic Affairs.

NEN-ISO 8402. (1994). *Quality control & quality assurance: terms and definitions*. NNI Delft. 3rd Edition.

NEN-ISO 9001. (1994). *Quality systems. Model for quality assurance by design, development, production, installation and servicing*. NNI. Delft. 3rd edition

NEN-ISO 9004. (1994). *Quality control and the elements of a quality system*. NNI. Delft. 3rd edition

NEN-ISO 10011. (1994). *Directive for the conduct of audits for quality systems*. NNI. Delft. 3rd edition

NEHEM. (1994). *MILKA: Een instrument voor de integratie van zorgsystemen voor kwaliteit en arbeidsomstandigheden. Handleiding aangevuld met ervaringen van 7 bedrijven.* (MILKA: an instrument for the integration of management systems for quality and working conditions. Guidance notes accompanied by the experience of 7 companies). NEHEM. 'S Hertogenbosch.

Norwegian Petroleum Directorate (1985). Regulations concerning the licensee's internal control in petroleum activities on the Norwegian Continental Shelf with comments.

v.d. Poel I., Hale A.R. & Goossens L.H.J. (2001). *Safety management in the Dutch oil and gas industry: the effect on the technological regime*. .... In press.

Robens. Lord. (1972). *Safety and health at work: report of the committee*. HMSO. London.

SodM. (1994). *De Nieuwe Koers* (The New Course). Rijswijk. Staatstoezicht op de Mijnen

SodM. (1996). *The health and safety report: a report to industry*. Rijswijk. Staatstoezicht op de Mijnen.

SodM. (1999). *Organisatieontwikkeling Staatstoezicht op de Mijnen* (Organisational development State Supervisoin of Mines. Rijswijk. Staatstoezicht op de Mijnen.

SodM. (2000). *Organisatie en Formatie Rapport*. (Organisation and training report) Rijswijk. Staatstoezicht op de Mijnen.

SSVV. (1997). *VCA: Safety checklist for contractors*. Leidschendam. SSVV.

Visser J.P. (1998). Developments in HSE management in oil and gas exploration and production. in Hale A.R. & Baram M. (Eds.) *Safety management: the challenge of organisational change*. Pergamon. Oxford.

Westrum R. (1998). Organisational and intra-organisational thought. World Bank Conference on Safety Control & Risk Management, October 1988.

Westrum R. (1991). Cultures with requisite imagination. In: Wise J., Stager P. & Hopkin J. (Eds.) *Verification and validation in complex man-machine systems*. Springer. New York.

Zwetsloot. G. (1994). *Joint management of working conditions, environment and quality*. Dutch Institute of Working Conditions. Amsterdam.

The regulation of rail safety is particularly topical given recent major accidents in the UK and in Germany. We are fortunate, therefore, to have contributions on this topic from both countries in this volume. The rail industries in both countries have experienced the move from prescriptive to goal setting/safety case legislation, but what has made this particularly controversial in the railway context that it has occurred concurrently with privatisation. This has raised a question in the minds of many as to whether these changes in ownership and regulatory arrangements are responsible for the rail disasters in these two countries.

Maidment's view is that in the UK there is no evidence that privatisation or the goal setting framework have caused a decline in rail safety. He argues however that the highly politicised context of rail safety has put rail regulation under intensive scrutiny, that the public increasingly expects accident free operation no matter what the cost, and that recent accidents have undermined trust in both the railways and there regulators. He believes that as a result the balance in the UK will tilt towards more prescriptive requirements being imposed on rail industry operators.

7

THE DEVELOPMENT OF SAFETY REGULATION IN THE RAIL INDUSTRY

David Maidment

BACKGROUND IN THE U.K.

The first public railway in the world (the Stockton & Darlington Railway) opened in 1825. At the opening of the Liverpool and Manchester Railway in 1830, the first accident to a member of the public caused by the new form of transport took place -unfortunately or otherwise, depending on your standpoint, it was the death of the MP and Secretary of the Board of Trade (responsible for the new railway industry development), William Huskisson, who was knocked down by a locomotive at the opening ceremony. Parliament therefore took an interest at a very early stage in the safety of the new form of public transport and safety regulation was in place by 1841, with important legislation in 1871 setting up many of the powers placed with Her Majesty's Railway Inspectorate (HMRI) that are still in force.

This legislation included powers of inspection and approval, and also independent accident investigation over and above that conducted internally by the railway companies themselves. After any major accident involving injury or fatality to members of the public (and sometimes to railway employees also) an Inquiry would be held, causes established and recommendations produced for implementation by the railway companies. In early days these recommendations drove the progress in implementing safety systems which have stood the test of time, particularly in the areas of signalling principles and rolling stock construction and braking systems. At a time when many independent railway companies existed (up to 1923), HMRI was able to identify best practice and use this to press all companies to adopt these.

Not only was there pressure to adopt best technical practice, but the same era saw the development of company rule books incorporating general standards of acceptable behaviour and best operating practice. In 1923 the many railway companies were amalgamated into four large private enterprise organisations, each with their own rule book, and in 1948 they were nationalised as British Rail (BR). BR produced its definitive Rule Book in 1956 which has only been significantly altered in the 1990s.

The HMRI stood outside and independent of the railway companies and was a powerful influence on the private railways, publishing an annual report giving safety statistics and comparing the safety performance of railways from early 1900s onwards. The HMRI was primarily staffed by experienced military transport experts who had led railway operations in war torn Europe during both world wars. There was a corps of Railway Engineers which provided officers with such expertise until the 1960s and some Railway inspectors with such a background survived until the 1980s. They were steeped in the railway culture, knew the technology and operating rules in detail and had the tradition of military discipline and hierarchy. A similar culture existed throughout the former British Empire.

The disbandment of the military base for such expertise and the development of modern traction and electronic signalling systems in the 1960s brought about a change in the practice of railway regulation. The expertise no longer resided outside the industry but could now best be found within BR itself and as a result, HMRI began to recruit mid career railway engineers from within the organisation. Because of the difference in pay structures between the civil service and senior engineering posts within BR, this meant that innovation and leadership on safety issues transferred over the years to BR itself. This was consolidated by the development of research expertise at BR's Derby Railway Technical Centre.

Because of these issues, a closer working relationship between the regulator and railway management developed, with progress on safety matters being achieved through consultation and consensus, with little external legal pressure being applied. The railway management clearly welcomed this and claimed that the consultative approach was more successful in achieving change rather than reliance on a more confrontational style, backed by legal action, which was the culture in some other industries. This led by the 1980s to the criticism that the regulator was too close to the industry and that the relationship was too "cosy".

As long as safety performance was improving and there were no major train accidents to stir public concern, the drive to reform the regulator/railway management relationship was weak. From 1967 until 1987 the only train accidents in the UK involving significant loss of life were ascribed to causes beyond BR control or where their ability to control was perceived to be low. (Polmont, collision with cows on the line, causing derailment and 13 dead; Lockington level crossing collision, a van on the crossing causing the derailment of a DMU and 9 fatalities). Accident Inquiries concentrated on the immediate causes ("triggers") of the accident and seldom investigated system failures and recommendations were prescriptive in nature, usually requiring detailed changes to rules or engineering standards or maintenance practice.

CATALYSTS FOR CHANGE

A series of major accidents in the mid/late 1980s had a profound impact on public concern on safety issues. Major catastrophes in space (Challenger), on land (Chernobyl and Hillsborough football stadium) and on water (Zeebrugge, Piper Alpha and Thames Marchioness pleasure boat) were followed by train accidents in France (Gare de Lyon) and the UK (Kings Cross fire, London Underground, and the Clapham Junction train collision).

Secondly, public acceptance of transport accidents as an inevitable price of mobility had been replaced by an unpreparedness to tolerate major accidents without recrimination, blame and litigation. Although the companies which caused accidents were mainly in the firing line, safety regulators did not escape scrutiny and units such as the HMRI began to be challenged over the closeness of their relations with the industry regulated and the lack of prosecutions initiated by them.

The reviews of safety systems initiated by London Underground (LUL) and BR in 1988 and 1989 as a consequence of the Kings Cross and Clapham Junction events were not prompted by the HMRI which still maintained a prescriptive and reactive approach. In fact, BR and LUL safety management had to argue the merits of adopting a "risk management" approach following their studies into best practice in other UK industries. It was only on transfer of the UK HMRI from the Department of Transport to the Health & Safety Executive in 1990 that new blood was recruited into the Inspectorate which recognised the value of more proactive and less prescriptive systems, and which supported railway management in developing its new approaches to safety management.

The HMRI had responsibility for the approval of new railway infrastructure and had developed over many years detailed standards for the design, installation and maintenance of railway equipment and hardware. These were incorporated in a document known to all as the "Blue Book" which acted as "Bible" for railway designers both within BR and LUL and the railway supply industry. However, over time the document had become outdated and a revision was planned in the late 1980s to take on board technical standards changes but still in the traditional prescriptive format.

The industry expressed concern at this and persuaded the HMRI to review the whole concept, resulting in a joint industry/HMRI senior level working party preparing a set of overarching principles for consultation and then backing this with guidelines for about a dozen aspects of railway operation (e.g. track standards, signalling principles, structures, stations, rolling stock, light railways, heritage railways) (UK H&S Executive, 1997). An important element of the revision was that the 36 overarching principles - for the first time - set down not only what had to be achieved but also why, and the guidelines for implementation gave scope for alternative systems and methods, provided safety equal or better than the guideline for "best practice" could be demonstrated. This, at least in theory, enabled innovation from the newly privatised industry through the safety case process.

Development of a Goal-Setting Regulation Framework

In the early 1990s BR set the industry pace in developing a more proactive approach to safety management, learning from the airline, oil, chemical and nuclear industries, adopting a risk management approach, applying quality management systems to safety, developing near miss reporting systems and taking on board insights from Human Factors academic research. Priorities for safety investment and maintenance were guided by a combination of strategic risk management and cost benefit analysis and safety budgets were prepared and agreed with government. The "As Low As Reasonably Practicable" (ALARP) approach was adopted as outlined in the HSE publication "Tolerability of Risk in the Nuclear Industry" (1987, revised 1992).

BR saw the value in using the safety case approach that was publicised in Lord Cullen's Piper Alpha report and started pilot applications for major infrastructure investment schemes using consultants as advisers. Throughout this period BR kept HMRI involved through normal communication channels, but the work was essentially driven by BR (with similar LUL initiatives) without HMRI creative input. Then the government published its privatisation proposals and required the Health & Safety Commission (HSC) to prepare proposals for the regulation of rail safety in the fragmented and competitive structure that was being considered by the politicians.

The HMRI were delegated by the HSC to carry out the regulation study and they consulted the industry widely, especially BR's Safety Directorate. Three options were considered in the report - regulation directly by the HSE through the HMRI; self-regulation by the railway infrastructure controller (Railtrack) with audit and overall supervision by the HMRI; and the separation of Railtrack/BR safety expertise into a new Safety advisory and regulatory body, on the lines of UK's Civil Aviation Authority. The latter was rejected by the government at the time, presumably on cost grounds. As a result of high profile accidents since rail privatisation, a Judicial review led by Lord Cullen is now proposing the creation of such a body. In the interim period, Railtrack's Safety

& Standards Directorate has been formed into a separate and subsidiary Railtrack owned company called "Railway Safety".

Direct regulation by the HMRI was rejected in the report as the Inspectorate did not have the comprehensive expertise required and there could be conflicts of interest between its regulatory and accident investigation roles. There had already been cases when a judicial Accident Inquiry was required by government because of the close involvement of the HMRI in the development of the relevant safety management systems or regulations.

The report recommendations, subsequently backed and implemented by government, envisaged a "cascade" model whereby the safety of the network would be the responsibility of the infrastructure controller, Railtrack, which would have itself direct control over risks from infrastructure installation and maintenance and safe train working through the operation of the signalling system. Railtrack's Safety & Standards Directorate, its safety advisory body at arm's length from the commercial line responsibilities of Railtrack, would co-ordinate and supervise risk imported by the Train Operating Companies (TOCs) and bring together the planning and performance monitoring for the whole network.

The mechanism devised to carry this through would be the Safety Case, whereby Railtrack would present to the safety regulator (HMRI) its own safety case for the management of the infrastructure and also for its management of the safety co-ordination of the network. Railtrack, on the cascade principle, would then receive and evaluate safety cases from the TOCs and would accept them after consideration and discussion, passing on approved documents to the regulator for noting and holding as legal documents against which the regulator would hold the TOC accountable in the event of train or occupational accident.

Railtrack's Safety & Standards Directorate produced the first Railtrack Safety Case in 1994 and this has been reviewed and revised at regular intervals since, both when there was significant change of system or organisation and at specific timescales to check against gradual cumulative change over time. From 1994 Railtrack put in place three senior teams to receive and review TOC safety cases. These teams included expertise on safety management systems (SMS), operating and engineering standards, human resource planning and training and external input from Railtrack's insurance adviser. Reviews were undertaken by document scrutiny and management team presentations and interview and the process for each TOC could take in excess of six months from presentation to issue of a certificate of safety, entitling the TOC to obtain a train running licence if other conditions (such as financial health) were satisfied.

The Railtrack Safety Case and those approved by Railtrack were not overly prescriptive in nature, but included a description of the activities undertaken, risk assessments of the hazards identified, analysed to an appropriate degree of detail, and descriptions of the control measures in force to assure Railtrack and the HMRI that all risks were both "tolerable" and "ALARP". The Railtrack Safety Case included the criteria by which the TOC performances and Railtrack line would be judged and by which safety investment and maintenance decisions would be made. These were deliberately explicit so that all TOCs could be seen to be under the same obligations and could not use different safety investment rules for commercial advantage.

The HMRI had the power to reject Railtrack's Safety Case and to question Railtrack on issues within TOC safety cases that they had accepted. It is pertinent to state that as far as this author is aware, no TOC safety case issue accepted by Railtrack has ever been challenged by the regulator. However, the Safety Case has been used by HMRI to test for compliance after an accident, and as it is a legal document, to mount prosecutions on the basis of non-compliance. Railtrack Safety & Standards has been carrying out safety case compliance audits as part of its own safety case commitments to control risks imported by TOCs, and HMRI can sample audit the Railtrack audits as a check on their thoroughness.

In practice, the HMRI touch on Railtrack has been a light one, until the recent Southall and Ladbroke Grove accidents when one has sensed a degree of defensiveness on the part of the regulator. Most media criticism has, however, been picked up by Railtrack both on its own line infrastructure responsibilities and as controller of TOC standards (whilst signals passed at danger - SPADS - are TOC responsibilities, management systems for their investigation and corporate learning fall to Railtrack S&SD). The Safety Regulator role has been seen as a reactive one and there has been some political pressure to push the HMRI into a more proactive role, possibly because of a perceived wish to "punish' Railtrack or blame it because of suspected conflicts between its safety and commercial roles.

CHANGING ROLES OF RAILWAY SAFETY REGULATORS WORLDWIDE

The attention of other national rail systems was drawn to the changing BR safety scene both as a result of the publicity surrounding the Kings Cross fire and Clapham Junction collision and also the impending privatisation of the UK railway network. Through articles and conferences many other railways, especially those with strong historic links to the UK, began to consider the BR "risk management" approach and to reconsider their traditional prescriptive standards and rules - which were becoming increasingly unwieldy and constraining in fast changing technology and commercial contexts.

In some cases - Hong Kong, Holland, France, New Zealand - the national railway systems took the initiative. In other cases - Australia, Canada, South Africa - the railway safety regulator was the prime mover. All were working towards - with differing emphases - less prescription, more room for innovation, more proactivity, simpler rules and more flexible standards. All were interested in the UK ALARP approach and the related cost benefit studies for assessing safety investment and budgets although there was scepticism among railway managers as to how realistic it would be in their own countries and in their legal systems.

In 1994 the South African Government instituted conversations between the nationalised transport operators, the civil service and rail safety regulators or advisers from the USA (Federal Railroad Authority), the European Union (UIC) and the UK (BR). As their intention was to move from state control to privatised transport sectors, the South African authorities were keen to learn from other regulatory sources. The FRA representative was very frank in advising the conference against the degree of rule and standard prescription laid down by the FRA, which he considered to be very bureaucratic and stultifying of innovation and initiative, but valid in his country only because of the large number of USA organisations many of which were too small to have competent safety expertise sufficient to manage a less prescriptive regime.

The UIC regime of interoperable norms or standards for EU railways was also seen as too heavy for the South African situation although the UK safety case approach caught the South African authorities in two minds as they were unsure that either the safety regulators (then spread over three government departments) or the railway operators were sufficiently mature to cope with far less prescription. A set of operating rules and standards for incorporation into South African law had been prepared for consideration by the South African railway management and a compromise was developed to allow more flexibility, taking the UK approach into some areas of operation. The South African railway managers were more inclined to the UK system than the other models presented to them and an impromptu risk assessment of the Soweto township suburban railway raised issues (such as on train murders and threatened violence to staff) that were very valid but went way outside the scope of safety management rules under consideration by those delegated to prepare appropriate legislation. The regulatory system has since advanced to require a "Rail Safety Charter" from each rail party involved, based on safety management policies, very similar to the safety case approach. (Villiers, 1999)

A recent safety study of the Irish railways (Iarnrod Eireann) conducted by consultants for the Irish Government included in its remit consideration of change in government safety regulation of the railways - again prompted by EU legislation and liberalisation of the rules concerning national rail systems. There was a small safety inspectorate within the Transport Ministry, but this had few legal powers and relied on goodwill and relationships with Irish rail managers. It was also severely restricted by its lack of resources and lack of depth of expertise that could reside in only a couple of posts. This degree of co-operation was now being tested by Freedom of Information Acts passed by the Irish Parliament, which had resulted in a drying up of safety and accident information flowing freely from railway to regulator for fear of the implications in such data being freely available. The main and legitimate concern was that frankness of reporting could be inhibited by the fear of recrimination from the media and public sources which railway management could not control.

In Hong Kong, the main mass transport railway (MTRC) had long taken the initiative in developing safety management systems and had agreed with the regulator that they would commission an independent and external wide-ranging safety review every three years that would report to top management and would be available to the Hong Kong Inspectorate. Until recently, the Hong Kong government had utilised a senior member of the UK HMRI with its UK based pre-1994 prescriptive approach. However, the two Hong Kong railways, MTRC and KCRC, had both adopted the UK goal-setting and risk management approach following the acquisition of UK safety consultants.

Current Issues of Concern to Rail Regulators, Governments and Rail Companies

Heightened public concern about rail safety and media exploitation of this, political reactions to this mood, fears about the effect of privatisations (threatened or implemented), especially the possibility of companies putting profits before safety, have all led to calls for safety regulatory regimes which are more proactive, confrontational and willing to use the legal process much more readily against both rail employees and companies. Safety managers and advisers within the industry see these trends as basically detrimental to rail safety maintenance and improvement as they are promoting a defensive safety culture which is more reactive to pressure, where unsafe acts and systems are not openly discussed because of external reaction, and safety budgets are distorted by pressure to spend to deal with the latest type of accident that has come under public scrutiny.

How should rail safety regulators deal with these conflicting pressures?

Independence / involvement

Some regulatory regimes (e.g. the UK system) seem to have started out as highly independent and powerful, become closer to railway management over the years, and are now being pushed to re-assert a more obvious and transparent independence again.

In 1990 the UK HMRI was transferred from the Department of Transport to the Health & Safety Executive where it became more open to the skills and practices of the regulatory regimes in other industries and could look for support from colleagues in the safety profession rather than rely on former colleagues in the rail industry.

Recent gut reaction of politicians to the Ladbroke Grove train collision near Paddington (October 1999) has been to focus public anger on the privatised infrastructure company (Railtrack) which had the responsibility placed on it by legislation of managing the network risks, with the HMRI holding more of an audit role. In response to the emotions it has stimulated, government spokesmen have made statements in which they state their intention of stripping Railtrack of its safety role, with the clear intention of involving the HMRI more directly or setting up an independent Safety Body for the industry similar to the UK Civil Aviation Authority (a proposal once recommended by the industry but rejected by the government at that time).

A few months later, the government, after receiving advice from independent safety experts, pulled back from its more extreme statements and indicated that it would only require some minor adjustments of the present regime, primarily the adoption by the HMRI of the role of approving the TOC safety cases. Already the HMRI has indicated that it will be seeking the assistance of Railtrack's Safety & Standards Directorate in carrying out this duty, as they have more expertise and resource in this process. Railtrack is also being required to put more clear water between its line company management and its Safety Directorate wing, setting up the latter as a company with its own Board - but this is to demonstrate greater independence from possible commercial pressures (in reality those who have looked for evidence of commercial influence have identified that most of the TOCs believe that Railtrack S&SD has too stringent a safety regime that is costly to them in time and finance and that it is insufficiently responsive to commercial realities). However, there are now signs that the government instituted judicial review of safety systems will look for a more independent body to supervise and lead rail safety and will bow to public and media opinion on the need to spend higher amounts of public money to prevent potentially catastrophic accidents than could be justified under the previously agreed ALARP/cost benefit criteria.

A genuine concern is that if the regulator takes too direct a responsibility for safety policy, setting the requirements in more prescriptive terms, the rail companies will react by implementing only what is explicitly required of them and will oppose improvement suggestions that are expensive, the regulator may lose this argument if its own expertise does not match that of the companies' in-house managers. This has recently been highlighted in a confidential report commissioned by a regulator in a British Commonwealth country, where a very proactive inspectorate (which had had recent experience in the national train company) was seen to be involving itself in day to day safety decision-taking, leading to strained relationships with the company and outright opposition by the company to safety measures being promulgated by the regulator.

A more direct role in developing safety standards and systems also means that the regulators' roles of system approval and audit, accident investigation and instigating legal penalties become compromised if the regulatory body has been implicated in decisions that are found to be less than satisfactory and have contributed to the accident. Who would then regulate the regulator?

Competence and expertise

Railways, and in particular national networks, can call on a large body of technical and operating expertise to devote to safety management, or to contribute to safety management, if their management has the will to do so.

Rail safety regulators rarely have a sufficiency of resources to develop the necessary comprehensiveness of expertise, the few staff employed tending to be experts in certain key areas only. ("The resultant lack of capacity to carry out proactive inspections has made it difficult for the Railway Inspecting Officer to gain adequate appreciation of the day to day safety management of the railway and resulted in a low awareness on the railway of a regulatory presence." (Welsby, 1999)

The National Department of Transport (South Africa) admits a lack of expertise in the development of a new Railway Safety Regulator (RSR) and states that it will rely on Spoornet (the South African national train operator) knowledge and experience, buying in some overseas expertise to test recommendations (to avoid other railway asset owners reacting negatively if Spoornet is identified as the sole driving force behind the RSR (Villiers, 1999).

The rapid development of technology has exacerbated this problem and many regulatory bodies are struggling with some of the latest computerised train control and signalling systems. In these circumstances the regulators have four options;

- buy in the expertise from industry if the requirement is ongoing and significant (but there is often a problem with the salary levels government funded regulators can offer);
- seek advice from other regulatory experts in other industries (as would be possible for the UK HMRI now working within the multi-industry HSE),
- commission consultant engineers and safety consultants to examine and develop subject areas in which their own expertise is lacking;
- develop a close relationship with the company being regulated and use their experts to provide the technical expertise, relying on the regulators' expertise in process matters of review, challenge, audit etc.

A further issue increasing in significance is the need for rail regulators to consider safety approvals given by other national regulators at a time of development of more interoperability of both equipment and train services between most countries in Western Europe. "For each country to tackle this task independently would be both onerous and wasteful since there may be competent authorities in other countries who have already conducted the exercise or for whom the technology is not new or novel. This raises the question of management of mutual acceptability of approval." (Welsby, 1999)

One way to increase the competence of the railway safety regulator has been suggested by Terry Atkinson, Manager Rail Safety, Land Transport Safety Authority of New Zealand:

"Because we have only one major railway in New Zealand there is a real need for us to have some basic performance measures available from worldwide sources to allow definition of tolerable safety performance of our railway and effectiveness of safety regulation. The principal purpose of our interest in international benchmarking is to allow us to gauge the safety of our industry relative to best practice standards so far as valid and reliable comparisons can be made, to provide a measurable basis of continuous improvement." (Atkinson, 1999)

To this end, the New Zealand regulator initiated a search of internet web sites to see if railway safety statistics might be represented on this medium. After a comprehensive search, only detailed data from the USA (Federal Railroad Administration), Canada (Transportation Safety Board) and the UK (HMRI) appeared to be available in that way. Even in these cases the definition of data collected was dubious and meaningful comparisons, difficult and possibly misleading.

Reactive/ proactive styles of regulation

There is much greater pressure in recent years for regulators to become more proactive, with the danger, developed above, that they may inadvertently undermine the initiatives and responsibilities of the line companies. As stated earlier, this can lead to a confusion of roles. However, there are different ways in which regulators can seek to be more proactive in meeting expectations increasingly placed upon them:

— they can become more prescriptive in setting standards, although this is very demanding in today's world of rapidly developing technology and in the requirements of commercial reality for greater flexibility,
— they can take a lead in commissioning research into safety systems or the safety implications of new technology, or ensure that there are competent bodies adequately funded to do this;
— they can become much more intrusive in daily operations exerting their right to inspect, check and audit at more frequent intervals;
— they can take stronger and earlier measures to enforce the implementation of recommendations flowing from rail accident inquiries, both internal and external;
— they can be more proactive in engaging the powers of the law, seeking prosecutions to push the industry to reform through fear of the legal consequences of not doing so.

What is clear is that if regulators are to play a more proactive role in promoting good safety as well as penalising shortfalls, they must have access to the expertise necessary to command the respect of the industry and be willing to develop and publish research and analysis that will give a rationale for actions they are advocating, able to withstand pressures from both companies and the public and their representatives where those pressures are based on unsubstantiated fears or commercial demands. This means that regulators might have to take cognisance of and apply the controversial techniques of safety cost benefit analysis and cost of life criteria to determine priorities - areas which they have left to companies to argue, whilst tacitly accepting some criteria in the past.

Supportive or confrontational approach?

The supportive approach is illustrated by this quotation from Linda Hoffman, Ontario Regional Director of Transport Canada (the regulatory authority):
"In the majority of cases, we work in partnership with the railways to identify and resolve problem areas. Rarely do we have to resort to more forceful enforcement action like formal warnings, "cease and desist" orders, or prosecution. Without the potential use of these sanctions, however, we would probably not be as effective or as quick in receiving voluntary corrective action from the railways." (Hoffman, 1999)

Recent safety management thinking has emphasised the importance of open and free discussion of safety problems, encouraging systems of "blame free" reporting of incidents and "near misses" to highlight potential accidents so that risks can be identified and controlled.

This development by a number of railway companies internally has been undermined by the tendency for the media and legal profession to seek retribution after any accident causing damage, injury or loss of life. A number of accident inquiries in the UK have been less than effective because of the early involvement of the Director of Public Prosecutions seeking action against one of the potential key witnesses. This issue was highlighted in the recent accident inquiry at Watford Junction when the driver's legal representative forbade him to answer questions from the investigating panel (it was a SPAD incident) for fear of incriminating himself before possible legal proceedings. It was also highlighted in the report of the chairman of the public inquiry on the train accident at Southall, when police investigations and securing of evidence delayed the railway industry learning the lessons for over two years.

A further conflict of interest has occurred in a number of countries where laws have been introduced allowing freedom of access to information held by government and public bodies. Whilst the intention is admirable, it has certainly led to a less open relationship between company and regulator with the fear that information disclosed to the regulator will become public knowledge. This will again cause individual operators and managers to seek to be less than frank for fear of media or public exposure and will cause companies to be hesitant about safety issues, especially the behaviour of people, unless there is proof that will stand a court's scrutiny as individuals involved in incidents will seek to defend themselves through legal representation because of the consequences of public exposure. ("Iarnrod Eireann, to whom the Freedom of Information Act does not apply, consider that their internal inquiry reports contain sensitive information the disclosure of which might compromise the effectiveness of the whole inquiry process by taking it into a quasi judicial mode ..... the forwarding of these reports to (the RIO) under established custom and practice has now ceased..." (Welsby, 1999)

Regulators have to decide what is, first and foremost, in the public interest and whether future safety is best enhanced by fear of exposure and prosecution or by persuasion using analytical rather than legal processes - or a judicious combination of the two. This can, of course, be influenced by the culture of the industry in any given country and point of time - whether it is a national system open to government control or a group of privatised companies, whether safety performance is improving and management is seen to be proactive or whether performance has stagnated and railway managements have become complacent.

Trust

Railways need to have a proactive search for safety enhancements that are rational, reflect known and identified risks and they need to develop a safety culture that delivers consistent implementation of the safety controls that maintain a safe system. Regulators need to ensure that railway companies fulfil these obligations and maintain their will to do this, despite all the resource and commercial pressures that might deflect their priorities.

These aims cannot be fulfilled without government and public trust that both the companies and the regulator are maintaining their vigilance and doing all that is reasonable to maintain and improve rail safety (whatever their tolerance might be of other transport systems where the acceptance of personal risk may be higher). When dramatic accidents occur, this trust is put under threat - especially as so often one serious accident seems to be followed by one or two other accidents in quick succession.

The UK Chief Railway Inspecting Officer, Vic Coleman, has raised the issue of managing and regulating safety at a time of growth of the industry since privatisation which increases exposure to risk both in terms of probability and consequence. ("The aim, in risk management terms, is to neutralise as many risk factors as possible and, where they cannot be so neutralised, to mitigate the adverse effects. If one can keep overall risk exposure steady in a growing industry, individual risk will effectively fall. This needs to be the minimum goal - in railways, as in aviation, the public sees the gross harm (in overall numbers of casualties or incidents) rather than individual risk exposure being the prime concern, so growing railways need to do better!" (Coleman, 1999)

On this issue, despite several major train accidents on Indian Railways in recent years, where fatalities have exceeded 100 (including two at 330+ and 280+ deaths), because of the sheer number of passenger journeys in India their safety performance statistically is better than that of Western European railways! However, trust in the safety of Indian Railways has fallen and after the last major accident a recently appointed Railway Minister felt obliged to resign.

Such trust can only be maintained if rail companies and the regulator are communicating about safety performance on a regular basis and are sharing the dilemmas and options for improvement in a transparent and open way. When other political agendas arise (such as controversies surrounding rail privatisation) it will become difficult to maintain objective views on safety performance as the various protagonists will seek evidence to support their preconceived views. It is of interest, however, that despite recent public and media outcry in the UK following serious train accidents at Southall and Ladbroke Grove within three or four years of privatisation, rational views on safety priorities and what is reasonable to invest, are beginning to re-assert themselves despite initial emotional reactions that threw such considerations out of the window. As BR and its successors had worked for a number of years to communicate its safety policies through publications like its annual Safety Plan, it might be postulated that such communication policies limited the damage to trust caused by the accidents. Only time will tell whether this is the case. (The rail accident at Hatfield in October 2000, caused by a broken rail, has once again put the privatised Railtrack under extreme pressure and the balance is now tilting towards external imposition of measures despite their cost, because of public lack of trust in Railtrack.)

CONCLUSION

Pressures on regulators to react to public, media and government concerns and to retain the confidence of the industries being regulated provide a difficult balancing challenge to regulators. The key word is balance - on each of the regulatory issues there is a balance between the extremes, which regulators may need to adjust from time to time to respond to external pressures or react to rail industry safety performance trends and attitudes.

It is suggested that regulators need to be aware of their stance on each of the potential "conflict" areas and that the use of such performance measuring techniques as the "Balanced Scorecard" approach which is used by the Hong Kong Mass Transit Rail Corporation (MTRC) to measure its performance on company objectives, and by France's Paris suburban network (RATP) for safety performance monitoring, could be beneficial. The scorecard can show schematically the trend towards reaction/proaction; independence/involvement; internal expertise/bought in expertise; persuasion/confrontation, etc. and, when drawn as is the convention, in circular format, this can demonstrate very easily the degree of balance maintained between the elements and where significant variations to the balance are occurring, the justification for which can then be examined. Such a technique could be used not only by the regulator, but also by the companies regulated and representatives of government and the travelling public, to both correct any bias of the regulator, and to highlight real or perceived difference of views by stakeholders in the industry.

It is pertinent that the UK Government has widened the remit of Lord Cullen, appointed to chair the Ladbroke Grove accident inquiry, to include a review of the UK rail safety regulatory regime and also that the EU has commissioned an exhaustive study of the various railway regulatory regimes in Member States. (Welsby, 1999) It may well be that the balance of regulatory policy is shifting to cope with the increasing organisational and ownership changes within the railway industry.

One railway regulator summed up his presentation to a recent International Rail Safety Conference thus:

"In general terms a move to a requirement for the developer/operator to demonstrate compliance/competence in terms of international standards and industry best practice appears the sensible choice as it affords the best use of scarce resources, offers the flexibility to accommodate industry and regulatory change and follows the philosophy already adopted in other sectors both at home and abroad. It also has the significant advantage of requiring the railway to establish a rigorous and auditable process of risk assessment and mitigation. While however an associated inspectorate might have the technical capacity to investigate railway accidents, the probability is that in the interests of independence and transparency this will have to be a separate function best provided for in a single cross-modal body." (Welsby, 1999)

REFERENCES

Atkinson, Terry (1999) *Railway Safety Regulation in relation to Tolerable Risk and Best Practice Benchmarking*, paper presented at the "International Rail Safety Conference", Banff, Canada, 19-22nd October 1999.

Coleman, Vic (1999). *The Safety Implications of Growth in the Railway Industry*, paper presented at the "International Rail Safety Conference", Banff, Canada, 19-22nd October 1999.

European Union, (1998). General invitation to tender VII/B2/35-98.

Hoffman, Linda (1999) *Canada's Role in Regulating Railway Safety: A Field Perspective*, paper presented at the "International Rail Safety Conference", Banff, Canada, 19-22nd October 1999.

UK Health & Safety Executive, (1997). *Railway Safety Principles and Guidance*.

Villiers, Johan de (1999) *Developing a Rail Safety Regulator for South Africa*, paper presented at the "International Rail Safety Conference", Banff, Canada, 19-22nd October 1999.

Welsby, John (1999) *New Regulatory Framework in Ireland*, paper presented at the "International Rail Safety Conference", Banff, Canada, 19-22nd October 1999.

Becker's is the second contribution on rail safety, this time from Germany. He describes the move towards goal directed regulation and privatisation in this industry and he raises very directly the question of whether this has had a deleterious effect on safety.

A major part of Becker's paper is a discussion of the ICE high-speed train crash near Eschede in 1998 which killed 102 people. He explores the causes of this crash and shows how the processes of privatisation and de-regulation contributed to this outcome.

He argues, further, that the new regime requires that companies develop strategies for organisational learning and he points out that inspectors cannot expect to have access to the information which companies collect for this purpose, lest this impede open internal communication.

He notes that the change has meant that inspectors spend less time in the field and more time on paper work and that this increases the risk that they will lose touch with what is happening in the field. He suggests that there is a need to develop indicators of how well a company is managing safety, which will give inspectors advance warning if safety is deteriorating, prior to the occurrence of any major accident.

8

Towards Goal-Directed Regulation in a Competitive World: Do We Underestimate the Risk of Changes in the Regulatory System?

Gerhard Becker

Introduction

The transition from rule-based to goal-directed regulation seems to offer great advantages, because it permits us to keep pace with the accelerating progress of technology in many areas. Faster adaptation of safety requirements to new developments appears possible, if industry bears responsibility for the detailed safety requirements, and the authorities monitor the relevant safety processes in the companies and their compatibility with general safety goals.

Working or industrial safety has been an area where German regulations, based on European directives, have followed the above trend, by the passing of the Industrial Safety Act of 1996. With this Act industry's responsibility has been strengthened, and the task of the state authorities has been shifted, from inspecting compliance with prescriptive rules towards checking documents regarding risk analysis and the resulting action taken by companies. This raises some questions with respect to the means available for sustaining the adequate feed-forward inspection necessary for preventing major accidents.

Some observations in German Rail as well in the nuclear industry confirm, in the author's view, that the shift towards more autonomy in safety matters to industry involves risks which should be discussed. An analysis of the factors contributing to the German ICE train accident at Eschede has produced the finding that the responsibility for permission to use a new wheel type for high-speed trains was changed during the licensing process. Due to the privatisation and reorganisation of German Rail, the Railway Inspectorate (Eisenbahnbundesamt - EBA) was empowered to give approvals for technical equipment. This body followed the principle of goal-oriented regulation, and monitored the process of development of the new wheel. Recent findings indicate that permission to use the wheel type was based on the manufacturer's design calculations and laboratory results only. Cost reduction seems to have been an important argument for the decision to use this wheel.

Observations in nuclear power plants indicate that managers who are under pressure from their shareholders, due to the more competitive market for electricity (as in other markets), are in danger of delegating more work to their subordinates than the latter can handle, if they comply with the procedures designed to ensure safety in their plants. Some examples are detailed in this paper.
In view of these observations, it is concluded that the transfer of more and more safety responsibility to managers in industrial settings who are subject to cost-reduction programs might need to be backed up with a more sophisticated approach towards independent audits and inspections.

Deregulation and Working Safety

Recent developments in safety regulation, towards greater responsibility within industry, obviously mirror a trend which regulation is following in Western Europe. As a result of the international agreement to dismantle trade barriers between countries and to harmonise regulation, a reorganisation of European regulations and standards is taking place. The arguments driving this process of reorganisation are that it will keep pace with technological change, and permit better adaptation to the increased competition on the world-wide markets. The widely accepted method of dealing with these challenges is to deregulate and transfer responsibility for safety to industry, because regulation is felt generally to be too restrictive.

The question which I want to discuss here is whether these intentions, driven by the wish to facilitate commercial success, may not involve considerable risks in the area of safety regulation. These risks result from the commercial pressure which may adversely affect a self-regulated company. The deterioration in safety standards might be hidden for some time, and only become apparent after an accident.

As an example of recent regulation following the trends described above, I would like to cite the German Industrial Safety Act of 1996, which applies European directives to German legislation

(Bieneck, 2000). According to this Act, the employer bears responsibility for analysing the dangers in various workplaces, and for implementing adequate safety measures. The responsible persons in the company have to be named to the safety authorities. We have now observed that this legal responsibility can have serious consequences for top managers. There are examples where, in cases of accidents causing the death of a worker, the plant managers have been condemned to go to prison, because they were found guilty of instituting insufficient safety precautions.

Ideally, inspectorates aim to prevent accidents, rather than simply reacting to them. One way to achieve this goal has been for the inspectors of the state authority (Amt für Arbeitsschutz) to visit workplaces regularly. Under the terms of the new Industrial Safety Act, responsibility for surveillance of the requirements of this Act remains with the authority. Formally there seems to be no change from the situation in past. But discussions with state inspectors have shown that there has in fact been a remarkable change in their working practice. Because of the new requirements that employers identify, assess and control hazards, they spend most of their time now on evaluation of the documents regarding the dangers at various workplaces, the action taken by the company, the organisation of the company's safety etc. The number of plant visits has fallen drastically, so that the detection of safety precaution deficits during visits is now quite random. Observed safety deficits used to be reliable indicators of a company's safety practices, and the information was effective in helping to prevent accidents. The inspectors fear that the level of prevention actually instituted by a company cannot be judged simply by reviewing documents, as the main instrument of surveillance. Diminishing safety levels in a company may therefore go undetected for some time, until something serious happens.

The shift of responsibility for safety matters to company management has led in practice to a reduction in the independent and routine monitoring of safety matters. In my view, the documentation required by the new Industrial Safety Act seems to provide only a very poor indication of the state of an accident prevention system, as seen from outside a company. This suggests that our feed-forward inspection system of preventive safety measures, as established over the years, may have suffered, and I think it is necessary to think about new tools or indicators to compensate for this.

In the context of work safety, the European directives had a further effect on regulation in Germany. The system of regulation in this area consisted of a tight network of acts, ordinances and administrative by-laws, which very often referred to technical standards or accident-prevention rules. Most of the detailed safety requirements were clearly defined in these documents, which could contain both product-safety requirements on the one hand, and instructions for the use of equipment in the workplace on the other. Under the terms of the European Agreement for Free Trade, governments have agreed to harmonise the standards which require product safety standards to exclude regulations directed at the users or operators. On the other hand regulations concerning work safety and accident prevention must not affect the features of the equipment used (Ackers &

Lambert, 2000). The necessary revision of a large number of the standards and accident prevention requirements involves many different groups, and will take several years. Revision of the work safety regulations will also aim to achieve more transparency, the reduction of detailed conditions, and the enforcement of companies' responsibility (Bieneck, 2000, Waldeck, 2000). The responsibilities of all companies are clearly defined in Paragraph 3 (Basic duties of the employer) of the new Industrial Safety Act (Arbeitsschutzgesetz, 1996). The announcement of fines and imprisonment in the last paragraph of this act is intended to enforce the responsibility of companies (note that this type of enforcement will only be effective after the event). It is too early to assess the outcome of this ongoing process, but it seems worth considering what methods and tools are available to ensure that the preventive aspect of the old regulations are transferred to the new rules so as to permit safety deficiencies to be detected in good time.

THE ICE TRAIN CRASH AT ESCHEDE

The ICE accident on 3rd June 1998 near Eschede in Lower Saxony, where a train travelling at about 200 km/h (125 mph) crashed into a bridge, killing 102 people and injuring hundreds more, may serve as a tragic example to demonstrate that safety may be the loser against commercial forces.

It is interesting to see how many commentators were shocked that there was no safety unit independent of the German Rail (Deutsche Bahn, DB) to inspect the design of the wheels and define the test programme for wheel certification. In fact the German Rail Technical Centre (Bundesbahnzentralamt, BZA), a department within German Rail, was responsible for certification.

The case of German Rail has important implications for the deregulation debate since it has autonomous responsibility for safety. The definition of safety standards and testing for compliance has been the above department's responsibility for decades, and this never gave rise to discussion. What has changed since 1990 is that commercialisation has reached German Rail. Since the government decided to convert the state-owned company into a privately owned joint-stock company, commercial forces have acted on the company in the same way as on all other commercially operated enterprises.

But let us look at this in a little more detail. What I want to illustrate here is only one contributing factor to the event, and that is the certification of the new wheel type, which played an important role in the accident. Quite a short time after the accident, during the search for possible causes, one of the wheels became the focus of investigation. The tyre of this wheel was broken. It had come from the leading axle of the rear bogie of the first coach. Interviews with passengers brought to light that they had noticed an unusual noise and movement about six kilometres before the bridge. The train continued in this condition, until it reached a turn-off approximately 300 m from the bridge. Here the flange of the broken wheel caught in a guide, and derailed the coach to the right. A little more than 100 m further on was another turn-off. This caused the next coach to derail, which in turn forced coach 3 off the line. Coach 3 slewed far enough from a straight direction to hit and demolish a central support of the bridge. The train separated between the third and fourth coaches. The bridge did not collapse completely until after the fourth coach had passed under it. When it did fall, it collapsed on to the fifth coach and cut it in half. The remainder of the train piled into the wreckage like a concertina. Only the rear power unit remained reasonably intact.

Further investigation then focussed on the design of the wheel. An analysis of the adequacy of the wheel design and development was carried out by the Fraunhofer Institute for Structural Durability (Institut für Betriebsfestigkeit, LBF) in 1999 after the accident. The report was delivered to the public prosecutor's office in 1999. Some details have been published meanwhile in the press (Der Spiegel, 1999a). The findings indicate that mistakes had already been made during the process of design and monitoring of the wheel. The tests to achieve certification do not seem to conform to state-of-art technology. Furthermore German Rail did not use inspection systems able to detect cracks which might start on the inner side of the tyre. This type of failure is believed to have caused the tyre to break.

When the ICE train was launched in 1991, it was fitted with monobloc wheels which were forged in one piece. But the high speed resulted in high wear, causing vibration which was transmitted through the steel-spring suspension units. Customers' complaints about the noise and discomfort meant that the wheels had to be changed every 100,000 km. To overcome the problems and to avoid costly repairs, German Rail decided to switch to a new design of wheel, which included a separate "tyre" with a rubber strip to reduce the vibration. Similar wheels have been in use for about forty years, but mainly for trams.

Nevertheless the new wheel Type No. 064 passed the test programme, and received a certificate for use on high-speed trains on 31st August 1992. The rules for the wheel certification test had been developed by the BZA, the German Rail Technical Centre, in Minden. German Rail's responsibility for safety has a long tradition, and gave no reason for complaints previously. One could raise the question of whether the regulator did not pay enough attention to the process of safety surveillance, because of positive experience in the past with the responsibility of German Rail for safety. It was planned, when German Rail became a joint-stock company, to shift the task of safety regulation and certification to the EBA Railway Authority, a state body like the British Rail Inspectorate. But before discussing what can be learnt, let us look at some details of the certification process which have come to light in the meantime, even though full information on certification has not been published yet.

Investigation by the prosecutor has focussed on three aspects, since it seemed clear that a material defect did not cause the tyre breakage:

(1) the rule defined by German Rail as to how long a tyre should be used before the wheel set had to be changed
(2) which rules existed for the inspection of the wheels
(3) whether the inspection rules had been obeyed.

(1) The tyre on the wheel wears away during use like a tyre of a car. German Rail permitted tyres to be used with an original diameter of 920 mm until worn down to 858 mm, before the trailer had to be withdrawn. After the accident, independent experts performed computer calculations. They found that an 890 mm diameter seemed necessary for safe operation. Below this value the stresses steadily increase, until the tyre breaks. The wheel used in Eschede had been used for about 1.8 million kilometres, and the tyre was worn down to 862 mm.

(2) The tyre broke after a crack on the inner side had grown to a critical size. German Rail had no inspection device able to detect cracks on the tyre's inner surface. This equipment was necessary in the view of the experts. Their opinion was expressed after the event. It touches on a more general question about regulators' decisions on new technologies, when knowledge of the causal mechanisms of accidents is poor. We will come back to this question below.

(3) The train's wheel system had been tested with a wheel-set diagnosis system at stated intervals. This system tested the roundness of the wheel - how much the wheel deviates from the ideal form of a circle. The inspection procedure, which dated from 1994, limited the roundness deviation to 0.6 mm, and required exchange of the wheel set for deviations higher than that. It was found that trailers were allowed continue in operation with a deviation of 1.1 mm. Data from two earlier inspections had already revealed deviations of more than 0.6 mm. The relevance of the roundness deviation limit was not clear to the staff, and the rule was in fact not known to them. It was only discovered by investigators after a lengthy search.

As a result of these findings, the public prosecutors requested German Rail to provide the minutes of board meetings dealing with the phase of wheel development and production. (Der Spiegel, 1999b). In May 2000 the prosecutors began inquiries into the former head of German Rail's certification department at the Technical Centre, into one of his subordinates, and also into the head of the design and development department and an engineer from the manufacturing company (KStA, 2000). The persons involved in the development of the wheel are accused of neglecting proper consideration of the question of structural durability.

This is only a selection of factors contributing to the genesis of the accident, which again confirm that latent failures, as defined by Jim Reason (1990), play the dominant role in causing this kind of event. With respect to lessons which might be learned, I want to reflect on some of the issues (or latent failures) described above regarding the relationship between the regulator and companies which are responsible for their own safety. The first issue concerns possible complacency of a regulator - in this case was German Rail itself - which may not pay enough attention to the processes of safety surveillance in a company, because of positive experience in the past. In the example cited above, it is clear that too little regulatory attention was paid, in view of the deficits in the certification process and the violation of the selection criteria during the wheel tests. The difference between these two aspects is that the first deals with licensing, and the second with surveillance of wheel operation. The surveillance of the operation of new technologies is crucial, to permit knowledge to be gained of factors which might be unknown or cannot be judged sufficiently during the licensing process. Complacency can be seen to be a common-cause factor (or a precursor, to use Reason's term, 1990) undermining both the aspects mentioned above, and is therefore a very serious danger.

Complacency is a form of mental bias - that if things are going well, they will continue to do so. It is quite understandable that this happens, but it is a real threat to safety. Before transferring more and more responsibility for safety to companies, we should find ways of avoiding the risks which may result from complacency within companies, and even within the regulator.

Another aspect which I want to discuss a little further is linked to the experts' statement that inspection devices should have been used which were able to detect cracks on the inner surface of the tyre. This statement was presented *after* the accident, when the cause of the accident and the horrible outcome was known. Obviously the regulatory experts of German Rail decided that it was not necessary to use this testing equipment when they put the wheel into operation. The probability of an accident resulting from a crack extending from the inner surface of the tyre was judged to be very low.

Such crucial questions occur very often in the course of the development of new technologies. Should the product or process be refused entry to the marketplace, or not allowed to operate, because certain necessary equipment may not exist? This is a difficult decision for the regulator (and the company), because the judgement has to be based on theoretical considerations regarding the probability of an accident with no historical precedents. The example of German Rail may indicate that a better feed-forward information system may also be needed in more traditional technologies like the railways. A better exchange of information across the boundaries of different industries may be advisable. The safety case strategies in the chemical industry, or the probability safety assessments in the nuclear industry and space technology could offer ways to overcome the regulatory conundrum to some extent. But when comparing the ICE accident with the origins of the Challenger accident, the main problem seems to be more that the organisational processes in companies which are under pressure to reduce costs or to have success need to be checked effectively from outside. For that - and here we are back to the point made above - we need effective indicators for the regulator (or any other unit in charge of monitoring).

Only by using suitable diagnostic tools able to identify failures in good time can we keep complex technological systems safe. We have already discovered the need for early-warning and learning systems, which should include near misses and organisational solutions (Becker, et al. 1996, Wilpert et all. 1996). In view of the impossibility of foreseeing all potential accidents in advance, it is necessary to support any activity that will improve learning during operation, including learning about the organisational context. But we have seen how difficult it is to achieve consideration of organisational factors, even in industries where all those involved are convinced that they cannot afford any serious accident (Becker, 1999a). An outside regulatory authority cannot have full access to all the information handled within the organisation, because this might jeopardise internal corporate learning from organisational deficits. But the authority should have indicators that permit the diagnosis of deterioration in the internal safety system. These indicators could be the number and type of improvements derived from an internal learning system as described in annual reports. Comparing these results between companies, and combining them with other indicators (e.g. the number of reportable events) allows the regulator to give feedback to the licensee, and to discuss observations like those described by Walter in this volume regarding the nuclear industry. But experience with this new type of indicator has not been reported yet, nor is it clear if such examples are applicable to less strictly regulated industries, which form the main group striving for de- regulation. Thus there seems to be a great need to explore the possibilities of defining useful indicators by appropriate research, because the changes in regulation are already progressing very fast.

In the context of German Rail, safety regulation had not been changed, but the law for the privatisation of the state-owned company was in preparation, and commercial attitudes were being applied very vigorously in the early 1990s in preparation for changing the railway into a joint stock company by the end of 1993. This shift of direction allowed latent factors to grow undetected, and resulted finally in the worst train disaster, which has ever occurred in Germany.

During this period of commercialisation German Rail was still responsible for the certification process. The wheel for the high-speed train was certified in August 1992. The Railway Authority was formed in 1994 to act as a railway inspectorate and one can speculate that its earlier establishment might have helped to avoid the disaster. In my view such independent monitoring of safety decisions is an important factor, and ensures that safety arguments are given their full weight when competing with commercial interests. But the possibility of such an independent institution raising safety arguments depends on it having the tools available to obtain the relevant information. Routine visits to companies performed by experienced inspectors, who apply their experience when judging if the safety measures of the company comply with preventive safety regulations, can be seen as an effective, experience-based method compatible with compliance-based regulation. Such direct possibilities of surveillance cannot be used any longer, after goal-directed regulation has been implemented as described above, and I doubt if new tools are available which are suitable for the new challenges.

The EBA, which has been in charge of the licensing and surveillance of the railways in Germany since 1994, has oriented its strategy towards the ideas of deregulation and process monitoring. Consequently it bases its decisions on the result of tests performed by a certified laboratory. But the only laboratory that has been certified by the EBA up to now is the one operated by the manufacturer of the products. This laboratory cannot be seen as independent of tensions between commercial and safety considerations within the company. When following the process-oriented concept, the EBA has to evaluate which of the laboratory's activities sufficiently satisfy the safety goals. As I have stated above, I doubt that we have the tools available to discover hidden danger factors during the licensing and certification process, when surveillance is focussed only on the process. As one of our experienced quality managers has remarked about his field: "The process-oriented inspection of quality management systems says nothing about the quality of the product that the company produces". One can argue that this is also valid for safety, because we seem to use similar concepts.

INFORMATION FROM THE NUCLEAR SAFETY AREA

The regulation of the German market for electric power was changed by law some years ago. The suppliers of electric energy have lost their monopoly as the only sellers of electricity in defined areas in Germany. Instead the customers are free to select other suppliers, and the local power suppliers have to open their cable network for the transmission of external electric power. All the German utilities have started cost-reduction programmes, and are reorganising their companies. Even the nuclear power plants have cut back their staff by up to 30%. There is no doubt that the cost pressure has reached all areas of these organisations.

On the other hand the utilities and their regulators are discussing safety culture. In the context of this debate, both sides have the intention to strengthen the utilities' autonomy in areas where human behaviour is relevant, and to limit the authority's actions to the level of checking the processes in the company relevant to safety, instead of being prescriptive. Intervention with the aim of strengthening safety culture in some nuclear plants has given us the opportunity to gain insight into some factors which might become relevant to safety, as a consequence of the modified regulations mentioned above (Becker, 1999b). The factors relevant in this context may include:

- managers under cost and time pressure might delegate too many tasks to their subordinates, more then they can efficiently handle;
- owners of a company may be persuaded to buy products or to employ contractors at the lowest cost, which will have follow-up effects on safety;
- engineers / operators who report deficits in procedures several times, without experiencing any reaction, may tend to deviate from written procedures;

- workers complaining about unrealistic procedures in their instructions may tend to deviate from the procedures;
- insufficient precautions for the handling of radioactive waste may lead to unrealistic time schedules for the maintenance staff and cause time pressure;
- the time of engineers in the production department should be used for operating tasks and for communication with their team members, and not for writing reports defending the relevance of technical / safety investments as demanded by the commercial manager;
- contractors' work is often judged to be cheaper, because the task of preparing the workplace and clearing up afterwards is done by their own people, and this is usually not included in their time schedule;
- time pressure caused by management (or by their own ambition!) may result in hasty behaviour and cause safety problems.

The above factors may be identified in time, by learning from minor events within a company. The prerequisite for the detection of such weak points is a well-established learning system. Learning systems are still the exception rather than the rule in industry, and the information is usually only available for internal corporate use.

Authorities in charge of monitoring the safe behaviour of industry, and obliged to limit their activities to the evaluation of the relevant safety processes in companies, will have no access to the information available from the above learning systems. Any attempt to request information may disturb the internal learning process, because willingness to deliver information about your own mistakes or errors will be very low, if the information is not kept within the company. Nevertheless it seems necessary that authorities should be able to detect a decreasing safety performance well before serious incidents occur. Previously, when safety performance was measured by checking compliance with rules, a decreasing performance could be easily detected - provided the rules were designed to ensure safety. Now, the authorities must develop new measures of how well an organisation's learning system is working.

CONCLUSION

Changes in a system, including a regulatory system, can introduce risk. In such cases increased regulatory attention is needed, instead of reliance on good track records in the past. Many companies have been requested to introduce safety measures, which identify the risks associated with change, and to develop management of change procedures. Changes in the regulatory system have not been given similar attention in the past. We may need thorough studies of the impact of regulatory change on safety. Some of the open questions are already becoming apparent.

The transition from prescriptive regulation to goal-oriented regulation, with the new task of the regulator to observe the internal company safety procedures, raises the question as to how feed-forward, precautionary safety standards can be maintained, which have been developed over the years and documented in the growing set of prescriptive rules.

The transfer of more safety responsibility to industry and the creation of more management responsibility result in intervention by the authorities, after something has happened. The question is how to observe a decreasing level of precautions or increasing risks in a company from the position of the authority, before something has happened. The method of learning from minor events, e.g. by reporting on human contributions to near misses, is mainly applicable to internal corporate systems, and useful for keeping the company management informed about safety weaknesses in advance.

The extended use of goal-directed regulation and surveillance requires methods or indicators to be found which allow assessment from outside as to whether a company's safety processes are properly based on precautionary principles.

REFERENCES

Ackers, D. & Lambert, J. (2000): Ein Grundsatzpapier für den Bereich "Arbeitsschutz und Normung. (A basic paper for the area of "industrial safety and standardisation") Die BG, February 2000.

Arbeitschutzgesetz (1996): *Gesetz über die Durchführung von Maßnahmen des Arbeitsschutzes zur Verbesserung der Sicherheit und des Gesundheitsschutzes der Beschäftigten bei der Arbeit.* (Industrial Safety Act – Act for the realisation of measures of work protection for the improvement of safety and health protection of employees at work).

Becker, G.; Wilpert, B.; Miller, R.; Fahlbruch, B.; Fank, M.; Freitag, M.; Giesa, H.-G; Hoffmann, S. & Schleifer, L. (1996) : *Analyse der Ursachen von "menschlichem Fehlverhalten" beim Betrieb von Kernkraftwerken.* (Analysis of causes of 'human errors' in nuclear power plant operation.) BMU-1996-454. Dossenheim: Merkel. ISSN 0724-3316

Becker, G. (1999a): From theory to practice – on the difficulties of improving human-factors learning from events in an inhospitable environment. In: Misumi, J. Wilpert, B. & Miller, R.: *Nuclear safety: a human-factor perspective* (pp. 113-125). London: Taylor & Francis

Becker, G. (1999b): Multi-level training workshops - an approach to spreading the ideas of safety culture in organizations. NetWork Conference Bad Homburg

Bieneck, H.J. (2000): *Innovation und Prävention für die Zukunft der Arbeit.* (Innovation and prevention for the future of work). TÜ 41(2000)1/2, p.3

Der Spiegel (1999a): Brinkbäumer, K., Ludwig, U., Mascolo, G.: *Die deutsche Titanic* (The German Titanic) No. 21/1999, June 20

Der Spiegel (1999b): *Akten bei der Bahn beschlagnahmt* (Files at German Rail confiscated) No. 21/1999, June 20

LBF (1999): *Fraunhofer Institut für Betriebsfestigkeit, Gutachten zum ICE Unfall in Eschede* 1998 (Expertise on the ICE accident at Eschede 1998), quoted in Der Spiegel 21 /99, June 20

KStA (2000): Staatsanwalt nennt vier Beschuldigte (Public prosecutor names four accused). *Kölner Stadtanzeiger* (Cologne daily newspaper), May 18.

Reason, J. T. (1990): *Human Error.* Cambridge: University Press

Waldeck, D.(2000): *Effizienz von Vorschriften und Regeln im Arbeitschutz.* (Efficiency of regulations and rules for industrial safety). Die BG (2000) 2, February.

Wilpert, B.; Becker, G.; Fank, M.; Fahlbruch, B.; Freitag, M. & Giesa, H.G. (1996): *Weiterentwicklung der Erfassung und Auswertung von meldepflichtigen Vorkommnissen und sonstigen registrierten Ereignissen beim Betrieb von Kernkraftwerken hinsichtlich menschlichem Fehlverhalten.* (Improvement of reporting and evaluation of significant incidents and other registered events in nuclear power operations concerning human errors) BMU-1996-457. Dossenheim: Merkel. ISSN 0724-3316

The chapter by Walther is the first of three dealing with regulation of the nuclear industry. These three chapters provide the possibility to assess how the approach differs in three European countries to what is a very high profile and international industry.

Walther describes the approach of the regulator to the Bavarian industry against the background of increasing competition and public concern for safety. He enumerates the integrated approach to technology, human and organisation and the tools available to the Bavarian inspectorate to perform their task. This is a description of a tight supervision, notably of the technical and human factors aspects, with the organisational issues being left largely to the power plants. Extensive use is made of risk analysis tools, particularly to prioritise actions and of detailed technical rules.

9

THE CHALLENGE TO SUPERVISION OF NUCLEAR POWER PLANTS UNDER CONDITIONS OF LIBERALIZATION AND GLOBALIZATION

Jürgen Walther

1 GENERAL

Since the beginning of its peaceful use, nuclear technology has played a crucial role in Bavaria's development to a modern state characterized by innovation and technology-awareness. The milestones in this development were the commissioning of the first research reactor in Munich, the test reactor in Kahl, the first commercially operated nuclear power station in Gundremmingen and, last but not least, the modern Konvoi nuclear power plant Isar 2. Nevertheless, this development has never been determined by blind and unquestioning faith in technology. Rather, a closer inspection of this development reveals that plant safety and thus a responsible approach to nuclear energy have always been indispensable in Bavaria. This becomes most evident in the "curriculum vitae" of the nuclear power station Gundremmingen Block A. For this plant, the Bavarian nuclear supervisory body had to demand such a volume of comprehensive safety-oriented retrofitting measures that the operator finally preferred to decommission Block A and to invest in the new construction of blocks B and C in Gundremmingen.

Five nuclear power plants with a total gross electric power output of approximately 6400 MW generate approximately 70% of the electricity consumed in Bavaria. Thus, in Bavaria, the portion of nuclear energy in electric power generation is roughly twice as high as in the rest of the Federal Republic of Germany.

The Bavarian State Government is convinced that nuclear energy utilization cannot be dispensed with in the foreseeable future. However, the basis for the assumption of the responsibility for this technology is that nuclear power stations are operated with the utmost safety.

Currently, we are experiencing predatory competition in the electricity market that even some months ago was still inconceivable. In order to ensure competitiveness of nuclear power stations in this changed environment, too, plant cost-effectiveness must be constantly monitored. The key factor in ensuring the cost-effectiveness of nuclear power stations is providing evidence of their safety, since without such evidence nuclear power stations are not accepted by the general public. In this context, we must strive to further enhance the already high level of safety in nuclear power stations without further increases in costs. This is the only way to prevent the accusation being made by the general public that in nuclear power stations "cheap electricity" is produced at the cost of plant safety.

For this reason, it is one of the objectives of the Bavarian nuclear supervisory body to optimize the enforcement of the German Atomic Energy Act by increasing effectiveness and focusing on the essential issues, so as to resolve the supposed conflict of interest between the requirement of unlimited safety, on the one hand, and cost-effectiveness, on the other hand.

2 Legal Framework of Supervision

In Germany, the peaceful utilization of nuclear energy and protection against its risks is regulated in general by the Atomic Energy Act (AtG). This law also provides the legal framework for the licensing and state supervision of nuclear power stations but does not define enforcement thereof down to the last detail.

The general provisions contained in the Atomic Energy Act are outlined in more detail in legal regulations pertaining to, amongst other things, radiation protection, criteria for plant-specific licensee event reports, and nuclear insurance cover.

Further detailed requirements and measures intended to guarantee the safe operation of nuclear power stations in line with the recognized rules of science and technology are specified in the comprehensive, constantly updated standards and guidelines of the Federal Republic of Germany.

The majority of these regulations refer to technical aspects. In addition, however, requirements and measures directly and indirectly connected with aspects of organization and human factors are also defined in detail.

In the past, the preparation of these standards and guidelines was characterized by deterministic approaches developed on the basis of operational experience in conventional plant engineering. Striving to ensure maximum safety, these deterministic approaches, as a rule, started from conservative safety assumptions and were oriented to what is technically feasible and not to safety-related requirements (e.g. leak-rate tests, containment, Castor contamination).

In spite of this very comprehensive set of standards and guidelines, it must be stated that not every detail of implementation and enforcement can be defined and that it has never been our intention to do so. Over-regulation would not make sufficient demands on the responsibility of the licensee and all the other parties involved, and, in the end, would jeopardize the aspired goal of enhanced safety.

On the other hand, permitting such leeway calls for effective and appropriate administrative procedures. Law enforcement and implementation characterized by ideology or opinions – even a policy of "pinpricks" -- is, in comparison, purely and simply illegal.

Such a policy of pinpricks means not to support, but to disturb an effective procedure. E.g. those pinpricks may be obtained by an extreme restrictive interpretation of rules, so that an optimization progress becomes nearly impossible.

Our endeavors, in conjunction with the predictable marginal conditions of the nuclear enforcement and implementation procedure, aim at encouraging licensees to adapt their plants to technical progress and not simply to be satisfied with the guarantees contained in their operating licenses, which in Bavaria are all, without exception, legally binding and unlimited in time.
Implementation of the Atomic Energy Law must be oriented to the principle of "federal loyalty" in Bavaria too. This means that we work strictly according to law under the legally defined supervision of the Federal German Ministry of the Environment and according to its provisions.

Even a new Atomic Energy Act, which might stipulate a withdrawal from the nuclear energy program, cannot restrict state supervision. In such cases in particular, i.e. when a plant approaches its decommissioning date, the effective and responsible implementation of the law must be ensured, so that no reductions are made in the field of safety since "the remaining service life of the plant is only short anyhow".

Because there is only a limited number of officials the regulator entrusts independent expert organizations with the clarification of technical questions. These are different organizations because of the high amount of special problems.

One of the most important independent expert organizations for the Bavarian regulator is the TÜV Süddeutschland, which was active for the regulator from the very start of the peaceful use of nuclear energy in Germany. In the highly demanding fields of nuclear technology and radiation protection, in particular, the TÜV Süddeutschland has a unique pool of experience at its disposal.

3 EVALUATION OF PLANT SAFETY

Experience has shown that evaluation of plant safety on the basis of the aforementioned deterministic standards alone is not sufficient and no longer in line with the state of the art. The state of the art is characterized by deterministic and probabilistic criteria being applied equally to the evaluation of plant safety.

Today, experience from a total of more than cumulated 9,000 years of operation of nuclear power stations is available. At the same time, the tool of probabilistic safety analysis (PSA) has been brought to the point of application (cf. Rasmussen study 1978). Operational experience and the fully developed tool of probabilistic safety analysis allow quantitative assessment of plant safety or of the balance of a plant's safety concept.

The combination of deterministic and probabilistic procedures in the evaluation of plant safety has been realized in the "approach oriented to safety objectives". This method verifies plant safety via the achievement of so-called "safety objectives". The standard to be issued by the German Nuclear Technology Committee, "KTA 2000", intends to establish the framework conditions for such an approach oriented to safety objectives. In Bavaria, this approach has already been applied to all nuclear power stations during periodical safety inspections (PSI).

These periodical safety inspections are performed every 10 years and include the following elements:

- description of the plant's safety status
- probabilistic safety analysis (PSA)
- evaluation of the plant's safety status and the way it proves itself in operation

These reviews must be regarded as supplemental to the constant checks that are part of supervising the operation of nuclear power stations as regulated by the German Nuclear Legislation.

PSI are carried out by the operator of the plant and they are evaluated by the authority with the assistance of TÜV. Basically they take about two years, but there is an additional strength of the operator in the time before officially starting the PSI to get good conditions (unofficial strength), because each operator is interested in a good result of the PSI. The necessary effort for the official part of the PSI is about 20 man-years in total.

With the results of probabilistic safety analyses (PSA), the achievement of safety objectives can be assessed on a quantitative basis and in line with their safety-related significance (Fig. 1)

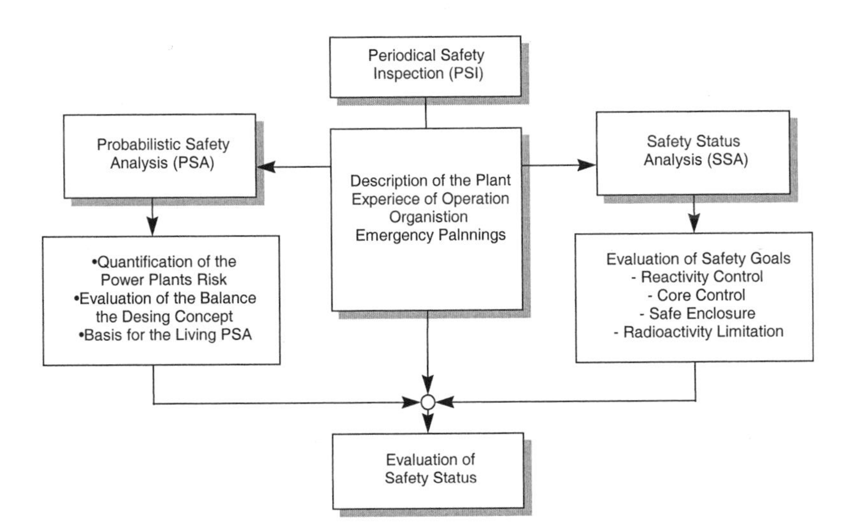

Figure 1: Procedure for the evaluation of safety status in periodic safety Inspections

The probabilistic information is an „additional" information to the deterministic information. So a difference in the evaluation is not possible. The probabilistic information helps to decide whether a deviation from a deterministic rule is acceptable.

Relating to Hopkins & Hale our evaluation of the plant safety is based on a goal-oriented legislation, which also uses technical prescriptions to verify the fulfillment of the safety goals.

4 INTERNATIONAL TRENDS

Recently, the economic constraints regarding cost-effective electricity generation have increased fundamentally worldwide. Likewise, the climate-related targets agreed in Kyoto show that, in the larger industrial nations, effective reduction in CO_2 emission can only be achieved if nuclear energy contributes essentially to electric power generation.

Equally uncontested worldwide is the fact that we can only continue to use nuclear power if its use is safe, reliable and cost-effective.

Again worldwide, the tool of probabilistic safety analyses in connection with deterministic safety verification has over the last years proved to be the instrument that can ensure effective use of nuclear energy.

The Vienna-based International Atomic Energy Association (IAEA) now promotes the application of probabilistic safety analyses in combination with operational experience and technical assessment as the tool which, through systematic selection and the establishment of priorities, allows focusing on certain areas and components which are significant for safe and reliable nuclear power plant operation.

Over the last years, the U.S. Nuclear Regulatory Commission has established the legal prerequisites of "Risk-Informed Regulation" to be applied to the procedure in line with the atomic energy legislation. This means, regulation which is concentrated on risk relevant items. Currently, these prerequisites are being put into practice in the "South Texas Project".
It became evident at international conferences that this procedure has been internationally recognized and thus, worldwide, represents the state of the art.

5 INTEGRAL APPROACH ORIENTED TO SAFETY OBJECTIVES AS A NECESSARY MEASURE TO SATISFY LEGAL REQUIREMENTS

Supervisory practice has confirmed the aforementioned experience again and again, i.e. that the state of the art in the field of safety cannot be represented by design-oriented deterministic standards alone, but, at the same time, must be oriented to the state of the art in technology and science. The safety-related objectives on which these standards have been based are therefore crucial. For this reason, design-oriented standards in general do not define how basic safety-related objectives have to be satisfied.

Our approach to the evaluation of plant safety both in the licensing and the supervisory procedure therefore focuses on the concept of safety objectives. In line with the above, the basic safety-related requirements which represent an appropriate safety level in operated nuclear power plants are oriented to the following Safety Objectives:

- monitoring and limitation of reactivity;
- cooling of fuel elements;
- containment of radioactive substances;
- restriction of exposure to radiation.

The approach is oriented to the recommendations made by the Reactor Safety Commission and takes into consideration the basic requirements of the Atomic Energy Act, the safety criteria defined by the Federal Ministry of the Interior, and the Reactor Safety Commission guidelines for pressurized and boiling water reactors (Fig. 2). As mentioned before, these requirements are to be integrated into the KTA 2000 standard.

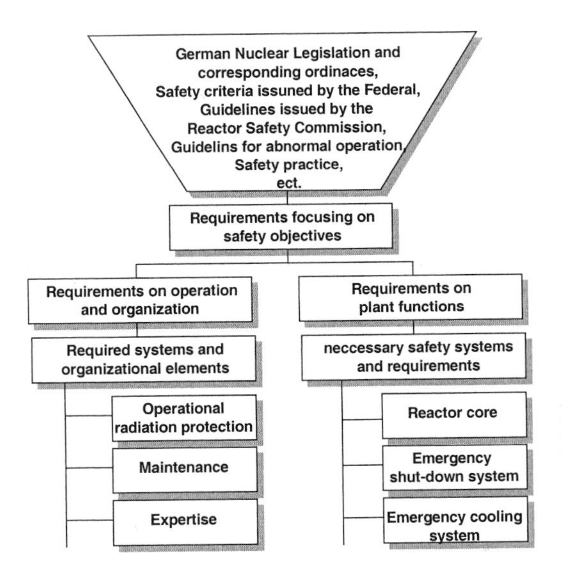

Figure 2:
Principle approach for converting safety relevant codes of Practice into requirements focusing on safety objectives

Of course the companies from the very beginning generally did not accept this procedure. Some sticks were necessary. An example is the support of the authority which can be given or not when the operator wants to carry out actions to improve reliability e.g. on line maintenance.

6 INTERACTION OF HUMAN FACTORS, TECHNOLOGY AND ORGANIZATION WITH REGARD TO THE SATISFACTION OF REQUIREMENTS ORIENTED TO SAFETY OBJECTIVES

Plant safety is always influenced by the following factors: man, technology and organization. These factors, in turn, are not independent of each other. Accordingly, evidence that safety objectives have been achieved must address all factors of influence. This can only be realized by means of an integral approach, which is the only way to ensure that interface problems between man, technology and organization are also sufficiently taken into account (Fig. 3).

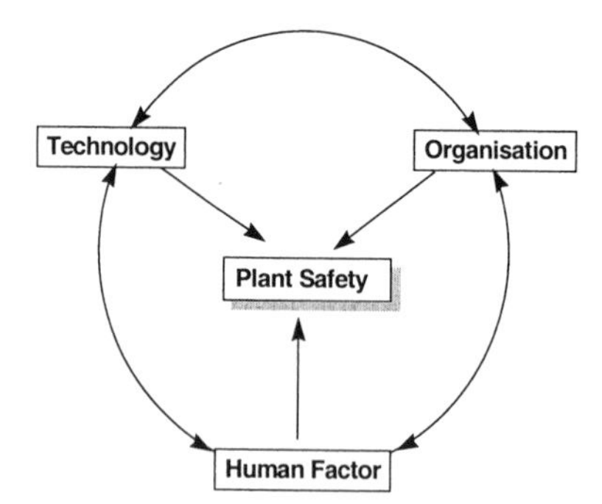

Figure 3: Factors influencing plant safety

In general, the following can be said concerning the significance of the individual factors:

6.1 Human factor

The feedback of experience from plant operation demonstrates that the relative significance of the human factor with respect to plant safety has increased. Due to constantly enhanced technology, the human factor has moved up in the ranking of influencing factors. Measures to enhance the reliability of human action thus may contribute significantly to an increase in the safety of nuclear power stations.

6.2 Technology

In the area of technology, a high level of safety has been achieved through ongoing optimization measures. Considerable gains in safety can now only be obtained through excessive technical efforts and at high costs. For this reason, activities aiming at an increase in plant safety are shifting more and more towards fine tuning.

In this area, we have to maintain the level of safety while adapting plants to the state of the art in science and technology within the scope of the Atomic Energy Law.

6.3 Organization

The structural and operational organization in nuclear power stations and the operating companies largely fall within the licensee's responsibility.

The effects of the structural organization on the safety of nuclear power stations are considerable. For instance, organization must ensure that responsibilities are clearly defined, a comprehensive flow of information is guaranteed and that the required measures can be implemented without delay.

In line with its significance, the structural organization of power stations is defined in safety specifications and thus requires approval by the supervisory bodies.

6.4 Interfaces between human factors - technology - organization

In addition to human factors, technology and organization, an integral approach must also take into consideration the interfaces between these elements. Practical experience has shown that interfaces are frequently not sufficiently taken into account. Gaps and misunderstandings in the flow of information at the organization/human factor interface as well as manuals and instructions (servicing, maintenance and repair) at the interface between human factor and technology that are not clear and unambiguous enough form the main causes of undesired events.

When an analysis is made of the overall system "human factor – technology -- organization", the interfaces in question must therefore also always be sufficiently addressed.

All in all, the integral approach towards verification of the achievement of safety objectives allows

- the focusing of measures on the key influencing variables,

- the orienting of evaluation to the safety-related objectives on which the standards are based,
- the evaluation of operational experience, and
- the incorporation of the state of the art in safety technology.

As far as there is no official connection between safety and economy, the key influencing variable include no performance criteria.

7 TOOLS IN THE BAVARIAN SUPERVISION PROCEDURE

Integral supervision oriented to safety objectives is aimed at considering all variables that may influence the safety of nuclear power stations. This should be done by means of a prioritized approach, i.e. the areas which are critical for power-station safety should be monitored more intensively than less important influencing variables. The principal approach is given in Fig. 4.

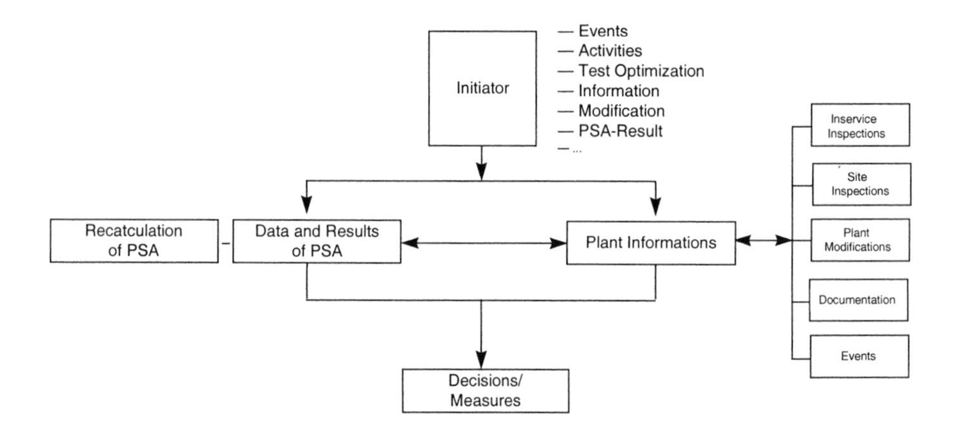

Figure 4: General Procedure for evaluation

Key elements in this integral monitoring system oriented to safety objectives are:

- information systems
- plant inspections (presence in plants)
- periodic tests and inspections
- emission and environmental monitoring
- review of personnel expertise

- human factor (HF) system
- assessment of changes in plant engineering and mode of operation
- periodic safety tests and inspections involving probabilistic methods

Altogether these tools allow putting step 1-5 of Fig.2 in Hopkins & Hale into action.

Characteristic of these elements is that they always incorporate all influences and impacts from the areas man, technology and organization. This ensures that problems at interfaces are also implicitly considered.

7.1 Information systems

Electronic data processing is an aid in the management of large quantities of data. Today, the performance of modern IT systems allows processing of almost unlimited information.

In a data collection system for plant monitoring and reliability analysis (Datenerfassungssystem für Anlagenüberwachung und Zuverlässigkeitsbetrachtungen (DAZ)) developed jointly by TÜV and the StMLU, and a plant and safety monitoring system for nuclear surveillance in line with the Atomic Energy Act (Anlagen- und Sicherheitsüberwachungssystem für die atomrechtliche Aufsicht (ASA)), a systematical and inter-plant approach to data recording and evaluation has been realized. All the information on the actual state of a power station, on essential changes, results of periodic tests and inspections together with safety analyses that are relevant to the safety of the power station, are collated and processed for evaluation. Excepted therefrom are data collated within the scope of the HF system that are subject to strict confidentiality.

Thus a system is available, which

- allows access to the safety relevant data of the respective power station at any time, and
- enables quick assessment of the significance of events through linking of information that evaluate safety.

7.2 Plant inspections

As a general supervisory measure annual plant inspections are performed by the supervisory body and the expert organization. These plant inspections include random checks to verify that the condition and the function of plant components and systems as well as the plant's mode of operation conform to the stipulated requirements and that, on the basis of the recognized state of the art in science and technology, there are no objections to the continued operation of the plant as a whole.

These plant inspections are based on the requirements outlined in the operating license which place an obligation on the licensee to have these tests and inspections carried out by officially authorized experts and subordinate authorities. In most cases TÜV Süddeutschland supports the authority with authorized experts. As already mentioned, TÜV Süddeutschland is a big independent expert organization which guarantees the necessary experience of its experts according to the need of the authority.

Plant inspections are divided up and distributed over the year in line with a schedule that has been coordinated with the licensee.

In addition, plant-inspections serve to focus surveillance measures on certain priorities that have gradually become known in the course of supervisory activities.

7.3 Periodic tests and inspections

Periodic tests and inspections defined to maintain reliable functioning of individual plant components or systems, e.g. testing the function of a valve or a unit, are carried out by the licensee at regular intervals in line with special testing and inspection regulations. Additionally, special periodic tests and inspections which address specifically safety-relevant plant components or systems are conducted. They too are carried out at defined intervals on the basis of requirements outlined in the licensing decision and in the presence of or sometimes directly by officially authorized third-party experts that have been called in (sometimes also by representatives of the State Agency for Environmental Protection). Among other things, these checks serve to verify that periodic tests and inspections which fall under the direct responsibility of the licensee are conducted as stipulated. The officially authorized expert describes any deficiencies detected during these checks in official records and sends them without delay to the authority. The operator subsequently must eliminate these shortcomings immediately.

In terms of an integral assessment of nuclear power stations, the objectives pursued within the scope of periodic testing and inspection must be compared against safety objectives and, if necessary, redefined. In this context, particular care must be taken to ensure that the deterministic evaluation of a test result is complemented by the probabilistic assessment of the condition of a system.

7.4 Monitoring of the environment and of emissions

Applying Article 48 of the German regulation on Radiation Protection, the surroundings of nuclear power stations are monitored in line with a predefined plan by determining the radioactivity in samples of air, water, soil and vegetation as well as the local amounts of radioactivity.

Using the remote nuclear reactor monitoring system, a series of important measurement values from nuclear power stations are continuously recorded and transmitted to the measurement network center at the State Agency for Environmental Protection. In this context, monitoring focuses in particular on the discharge and release of radioactive substances via the exhaust chimney and also takes into account the meteorological variables which are significant for the spreading and deposit-formation of radioactive substances.
Significant operational parameters for the recognition of a potentially increased discharge of radioactive substances at an early stage or parameters which supply information about the current operational condition of a plant also are transmitted via the remote nuclear reactor monitoring system.

7.5 Personnel expertise

Reviews of the expertise of employees in charge and other staff have proved excellent tools for supervisory monitoring. In addition to formal technical examinations for shift personnel, which are held with the participation of the StMLU, measures to maintain expertise and training courses for employees in charge and other staff form the decisive basis for safe technology management at nuclear power stations.

7.6 Monitoring of human factor system

Bavarian operators have introduced the human factor (HF) system to optimize impacts on human performance. This system is used, among other things, to collect national event reports, plant-specific licensee event reports (LER) and confidential information, and to analyze them for evidence pointing to any optimization potential.

The key factors influencing human action are the factual and human prerequisites of performance.

Within the scope of our monitoring system, we focus on the factual prerequisites of performance, since these are the prerequisites of human performance factors (Fig. 5).

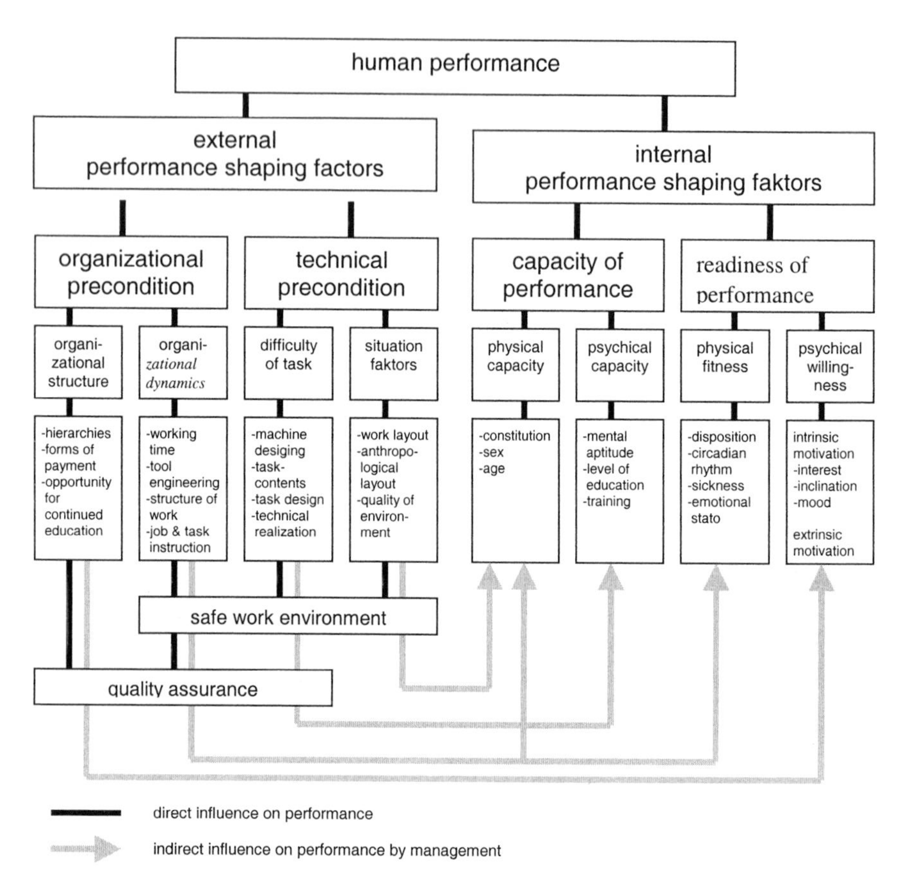

Figure 5: Human performance factors

Factual prerequisites of performance are largely determined by

- organizational conditions and
- technical conditions.

Organizational conditions. The majority of organizational conditions fall within the direct responsibility of licensees. In this context, monitoring is restricted as follows: the person who on behalf of the authority or expert organization conducts the regularly scheduled plant inspections watches out for indicators that make optimization of the organizational conditions appear feasible. These facts then are communicated to the licensee, who is directly responsible for processing the information and for developing optimization measures, if necessary.

Technical conditions. The optimization of technical conditions is carried out by the licensee within the scope of an HF concept. With the help of the HF concept, information revealing deficiencies in workplace design and surroundings are systematically recorded and ways of improvement developed therefrom. The following input is worked into the system:

- perceptions by the operating staff, including confidential information of employees;
- data from national event reports and plant-specific licensee-event reports;
- perceptions made by officially authorized experts during the conduct of periodic tests and inspections as well as plant inspections.

Special attention is paid to the interface between man and technology. Beyond the analysis of the man-machinery interface which is made by the licensee, conspicuous occurrences at the man-organization interface also are discussed in HF meetings at the end of the year (quality assurance in general, servicing and maintenance).

The plant-inspections conducted by TÜV are used to enter into the operational system perceptions which are collected on this occasion and which are relevant for the HF scheme. In turn, processing of the data resulting therefrom allows conclusions to be made with respect to system effectiveness. The general procedure is given in Fig. 6.

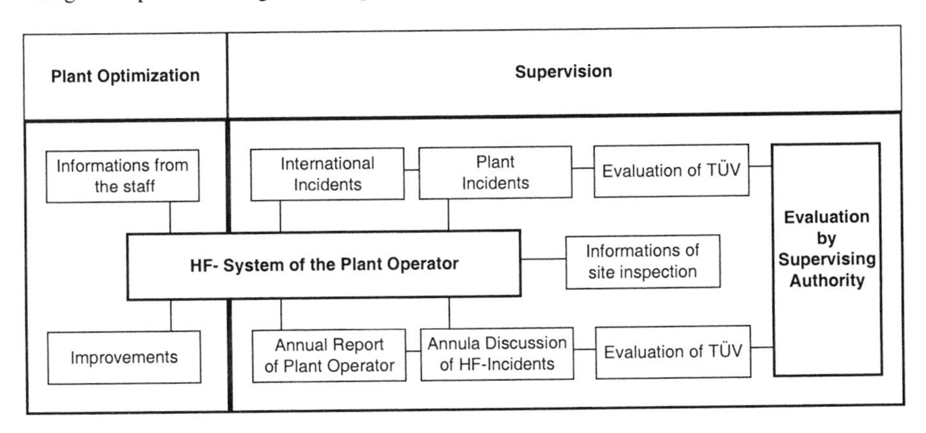

Figure 6: Integration of the HF-System

7.7 Assessment of changes

Major changes affecting a nuclear plant or its operation must be approved on the basis of Article 7 of the Atomic Energy Act.
In addition, however, minor changes in the plant may become necessary, for example, because, after several years of operation, individual components must be adapted to more recent developments in technology, system-related optimization is required or operational processes can be considerably enhanced.

Besides this, it may be mentioned that in Bavaria we do not actually differentiate between major changes, i.e. changes that need to be approved, and other changes not regulated by law. In both cases, the extent of the tests and inspections required is the same. Only the formal requirements outlined in the Atomic Energy Act and in the procedural regulations based on the Atomic Energy Act are exclusively applied to major changes. In return, procedural regulations defined by us which additionally have been embodied in requirements outlined in the operational licenses or operational manuals are applicable to minor changes.

In detail, evidence that the appropriate safety-related function, also in interaction with existing elements, and the necessary component reliability are ensured must be provided for every change that has been made in a plant. Thus the procedure applied in the case of changes allows constant adjustment of the technology deployed to the progressing state of the art in science and technology.

Beyond individual changes, the tests and inspections conducted provide a detailed picture of the technical condition of the plant as a whole.

7.8 Periodic safety inspections

Periodic Safety Inspections (PSI) which from their conception are carried out on the basis of an approach oriented to safety objectives are essential tools for integral plant safety evaluation. The main task of the PSI is to verify in how far the plant corresponds to the state of the art in safety technology.

Elements of the PSI are Safety Status Analysis (SSA) and Probabilistic Safety analysis (PSA). A security analysis to evaluate measures against external influences is not at the moment included in the scope of PSI.

8 INTEGRAL APPROACH TO SAFETY-ORIENTED IMPLEMENTATION

In addition to design-related and manufacturer requirements, safety-relevant functions are safeguarded through monitoring of operations, state-of-the-art testing, inspection and maintenance. In this context, the state of the art has been determined on the basis of general operational knowledge and experience after the occurrence of damage. In line with the above, these requirements generally reflect exclusively deterministic aspects and show only the starting points of a risk-oriented approach to determine the optimum measures to ensure the safety-relevant component function.

The integral approach in safety-oriented implementation allows deviations from the rigid requirements outlined in standards, for example, as far as the scope and frequency of component testing is concerned, without jeopardizing their basic objectives.

To apply the integral, safety-oriented approach, the extent to which components contribute to plant safety (probabilistic safety analysis) and the effectiveness of measures to ensure component function (deterministic approach) must be known. Thus the benefits offered by both the deterministic and probabilistic method are used to ensure the effectiveness of nuclear supervision (see also Fig. 4)

This procedure makes great demands on the qualifications of those involved. These people must possess comprehensive expertise and must not be afraid of making decisions, since – as experience has shown – all tolerated deviations from regulations and standards must always be justified in detail and defended in controversial discussions. In general this procedure is the implementation of step 3 in Fig. 2 of Hopkins and Hale.

8.1 Knowledge-based determination of the safety-oriented measures accompanied by full use of optimization potential

This procedure aims at adjusting the time and work needed to ensure the functioning of a component to the latter's safety-related significance and avoiding unnecessary monitoring, testing, inspection and maintenance. The time and work needed for safeguarding the functioning of a component thus becomes proportional to the component's safety-related significance.

Within the scope of PSI, a component's safety-related significance is determined in probabilistic safety analyses (PSA). In probabilistic safety analyses, systems are described by means of fault trees. These determine the function-related interrelationship between the individual components of a system. Fault-tree resolution i.e. the determination of a system's failure probability, is realized in fault-tree analysis. For this purpose, all combinations of component failures which may lead to system failure are determined in a first step. In a second step, these events are weighted by being assigned a probability of occurrence.

The probability of component-failure occurrence is largely determined by means of deterministic stress analyses which supply detailed information of the damage mechanisms of a specific component and their significance for the component's overall failure behavior. Thus the findings obtained in stress analysis and the results of fault-tree analysis allow knowledge-based determination of the optimum method to safeguard the function of a component.
Safety-oriented optimization of the concept will result in a balanced mix of monitoring, testing and inspection, and maintenance.

8.2 Evaluation of the safety-related status within the scope of Periodic Safety Inspections (PSI)

Within the scope of PSI, the safety-related status of a plant is evaluated integrally by means of PSI elements, i.e. Safety Status Analysis (SSA), Probabilistic Safety Analysis (PSA) and evaluation of performance in operational practice.

This invariably involves verification to see whether the aforementioned safety objectives have been adhered to. For this purpose, evidence must be provided (either through a deterministic and/or a probabilistic approach) that the requirements oriented to safety objectives have been satisfied. Satisfaction of the requirements oriented to safety objectives guarantees that the observance of higher level safety objectives is safeguarded.

Safety Status Analysis (SSA) provides a deterministic evaluation to verify that the requirements oriented to safety objectives have been satisfied. If this is not fully the case, operational experience gained in the plant itself and similar plants may also be used for evaluation. The extent of evidence required depends, amongst other things, on the frequency of occurrence of the respective events or their consequences. That is to say, probabilistic criteria determined within the scope of Probabilistic Safety Analysis (PSA) already play an important role in this area too.

Quantitative assessment of plant safety and the balanced nature of the plant concept taking into account all influencing factors is effected within the scope of Probabilistic Safety Analysis (PSA). The PSA includes human error/failure events as far as they are included in the operational manuals. Human actions described in the emergency manuals are not included.

9 RESULTS OF INTEGRAL SUPERVISION

Experience has shown that integral supervision provides benefits to both licensee and supervisory body. Essentially, these benefits express themselves in the safeguarding of knowledge and experience, and in optimized plant operation.

9.1 Safeguarding and use of knowledge and experience

Employee knowledge and experience are key prerequisites for the safe operation of nuclear power stations. However, knowledge and experience are not fixed quantities but constantly collected, expanded and updated in the course of plant operation and familiarity with the plant. At the same time, plant complexity requires highly specialized knowledge which might leave some gaps in the interfaces with other disciplines or as far as the consideration of marginal conditions is concerned.

Within the scope of the integral approach, all available knowledge, all plant data and operational experience are linked and made available to employees by means of the information system. Efficient evaluation procedures provide valuable aids for decision making concerning operational management as well as plant and process optimization.

The database is continually updated by the staff StMLU and TÜV and used in the daily decision making. So the information transfer to the staff in both the company and the regulatory body is guaranteed.
The loss of knowledge which may occur due to operating staff turnover, as well as the costs and efforts involved in training measures either to adapt operating staff to technical changes or to induct new employees is thus minimized.

9.2 Preventive maintenance

To prevent individual measures having to be carried out all at once, we support measures taken by licensees to carry out preventive maintenance during power operation (in-service preventive maintenance).

In this context, the introduction of in-service preventive maintenance invariably depends on the fact that integral plant safety is not affected thereby. This is to say, the loss in availability caused by measures of in-service preventive maintenance must be offset. The integral approach allows evidence to be supplied that integral plant safety is being adhered to. In the special case of in-service preventive maintenance, evidence is generally provided in the form of a mathematical approach through PSA.

9.3 Optimization of the periodic testing and inspection concept

In cooperation with licensees, a periodic testing and inspection technique which allows a shifting of testing and inspection activities to the phase of power operation or makes possible longer intervals between tests and inspections without compromising plant safety has already been developed. Suitable measures, such as consistent evaluation of operational information and application of progressive testing and inspection technology, have been developed for this purpose.

As already described for in-service preventive action, the same principle, i.e. integral plant safety must not be affected by changed or extended testing intervals, invariably applies to this case too. The integral approach allows licensees to provide the required evidence and additionally reveals possibilities for optimizing the periodic testing and inspection concept.

9.4 Safety status of nuclear power stations in Bavaria

PSIs have shown that, on the basis of solid design with changes implemented over the course of their service life as well as operational management, the safety of Bavarian plants represents the state of the art in safety technology. However, the integral approach applied to PSI also revealed that plant concepts can still be optimized in terms of safety technology. On the one hand, additional work and costs are incurred by such safety-related optimization, e.g. retrofitting measures, on the other hand, however, the monitoring concept may also be streamlined without affecting plant safety.

All in all, the status and the results of PSI in Bavaria demonstrate that the willingness of nuclear plant licensees to not only maintain but even increase the safety of their plants goes far beyond legal requirements. Last but not least, this may also be attributed to the framework conditions created by the Bavarian authorities, which especially emphasize the observance of safety objectives and further enhanced safety instead of insisting on point-by-point adherence to design-oriented standards.

10 SUMMARY

Economic demands placed on operators of nuclear power stations are increasing as liberalization and globalization progress. Even if, without any doubt, licensees consider the safety of their plants their first priority, they are still forced to further optimize plant operation.

Experience gained in the operation of Bavarian nuclear power stations has shown that our supervisory concept ensures the safety of nuclear power stations while at the same time revealing possible optimization measures. We are convinced that we have to pursue our adopted course further. When all is said and done, this will further enhance the safety of nuclear power stations without jeopardizing profitability.

Williams' chapter provides a parallel to Walther, describing the approach of the British regulator to the nuclear industry via its licensing approach. This places the emphasis on the licensee to explain and justify the measures taken and the organisation to keep them in place. This implements the safety case approach described in chapters such as that by Hopkins. Williams' chapter places much more emphasis on safety management issues, including the power of the Inspectorate to demand analysis of organisational change and its impact on safety before permission is granted for such change to take place. He uses the issue of new technology (largely software) development to illustrate the challenges of technological innovation for the regulator. In his discussion he raises issues of the need for the competence of the inspectorate to match that of the industry on such crucial new issues. He emphasises the role of the regulator in conducting research for and with the industry, as a way of keeping ahead of the game. He ends on the question of how to establish and retain the necessary level of trust to make the regulatory relationship work.

10

New Frontiers for Regulatory Interaction within the UK Nuclear Industry

Jeremy C. Williams

Regulation

The fundamental basis of UK nuclear regulation is that the licensee bears the sole and absolute responsibility for nuclear safety and this cannot be delegated to another party. Section 1 of the Nuclear Installations Act 1965 (as amended) specifies that no person shall 'use' a site unless a licence has been granted by the Health and Safety Executive (HSE). Section 3 specifies that a licence can be granted only to a corporate body. The Nuclear Installations Inspectorate (NII) interprets the user of a site as being the corporate body which is in day-to-day control of the site, process and activities, and whose staff manage the operation of the plant. This implies that there may therefore be circumstances where a contractor rather than their client may be held to be the user and hence require to be licensed (Taylor and Turton, 1998). NII will not grant two licences for the same site or grant one licence to two corporate bodies. This is because it is important that no doubt exists as to the identity of the corporate body having the responsibility for the safe operation of an installation (Taylor and Coatsworth, 1999).

Nuclear site licence holders are also covered by section 7 of the Nuclear Installations Act which imposes an absolute duty to prevent damage or injury in relation to nuclear matter and sets up a system of no fault, but limited liability. This means that the licensee is liable for any such injury or damage caused by an accident or by the fault or negligence of a contractor (Taylor, 2000).

Standard conditions are attached to site licences under the Nuclear Installations Act which, amongst other things, require UK nuclear licensees to make and implement adequate arrangements for the investigation and reporting of incidents, suitable training of all those on site who have responsibility for any operations which may affect safety, and for producing and assessing safety cases to justify safety during the design, construction, manufacture, commissioning, operation, modification and decommissioning phases of the installation (HSE, 1994). The licence conditions also require licensees to undertake regular and systematic examination, inspection, maintenance and testing of all plant which may affect safety and to ensure that all operations which may affect safety are carried out in accordance with written instructions.

These legal requirements place a duty on the licensee to apply good, modern practice with respect to engineering and management processes (HSE, 1996), and carry with them a responsibility for the application of sound human engineering principles and management practices. A new standard Licence Condition, number 36, imposes specific requirements for the Control of Organisational Change. This Licence Condition deals with organisational structures and resource levels and treats people as having the same importance for safety as plant or procedures (Furness, 1999a).

In addition to the Nuclear Installations Act, the Health and Safety at Work etc. Act 1974 (HSWA) places general duties on employers for the overall management of safety. In the last decade, significant additional requirements have been placed on duty holders via regulations made under the HSWA, such as the Management of Health and Safety at Work Regulations 1992 and the Construction (Design and Management) Regulations 1994. The HSWA employs inherently precautionary principles which are designed to ensure that those who create risks are responsible for protecting workers and the public from the consequences and that suitable controls are in place to address all significant hazards, irrespective of specific, situation-based, risk estimates.

The UK health and safety regulation system is risk-based, however, because the basic concept of the HSWA is to reduce risks so far as is reasonably practicable. This codifies the Robens Report view, that regulation based on sets of detailed rules was often unable to keep up with the pace of change in Industry. An approach which set goals, based on risk reduction, was preferred to cope with a changing world (Pape, 1996). 25 years of operational experience of the HSWA and 40 years' experience with the Nuclear Installations Act show that both sets of regulation work well, are highly cost-effective and allow more readily the licensing of nuclear facilities other than reactors.

REGULATORY INTERACTIONS

The primary aim of NII's assessment and inspection is to establish that the licensee retains and exerts effective control at all times. NII's approach to regulating the UK's nuclear industry is via its well tried and tested non-prescriptive nuclear licensing regime, which, as mentioned, involves granting licences to which standard conditions are attached in the interests of safety. It requires licensees to propose and establish adequate and integrated baselines for safety organisation, and mechanisms to address the impact of change on these baselines. It then inspects and assesses these baselines and mechanisms, and the outcomes of the licensees' activities and change processes (Weightman, 1999).

The role of the NII, therefore, is primarily a monitoring function, to test both the adequacy of arrangements and the degree of compliance with those arrangements. The NII also has "permissioning" powers which focus on proposed changes. Where appropriate, changes can be prevented by withholding permission. A licensee will not be given consent or approval for a safety-significant change until NII is satisfied that a good safety case has been made, and that they will adhere to it. In addition to these powers, NII also possesses intervention powers which can be used to enforce compliance. This is achieved via specifications, directions and improvement and prohibition notices, tools which are used rarely, given the basic permissioning concept.

NII's two main monitoring activities are the assessment of safety case submissions when licensees seek to change some particular aspect of their activities (plus periodic outage and other reviews); and inspection to verify that the safety case and other licence conditions are being complied with. A particular interest is taken in the licensee's processes for managing safety case changes, including the production of the case for change, review arrangements and formal endorsement by their Nuclear Safety Committee (Pape, 1996).

ORGANISATIONAL, MANAGEMENT AND HUMAN FACTORS

In a recent survey (Hollnagel, 1994) the contribution of human failure to the failure of hazardous systems was assessed as having increased over the past few decades by a factor of four. In the nuclear industry, even though incident rates are low, the proportion of incidents in which organisational, management and human factors can be seen to have played a part in the design, assessment, inspection, operation and decommissioning of nuclear plant appears to be increasing. The significance of this general observation is that, whilst the scope for technical improvement may be diminishing with time, it appears that major improvement may now have to come from areas over which licensees have day-to-day control but which they have not so far considered as being particularly amenable to further enhancement.

As a consequence of these observations, most regulators have conducted, or are in the process of conducting, research into the impact of organisational, management and human factors in the operations that they regulate. Emerging from this activity there is an apparent commonality of perspective across a wide range of cultures and regulatory systems. Researchers have identified a relatively small number of factors as being likely potential influencing factors, some of which appear to be amenable to a degree of modification. Most researchers report factors which include variants or developments of the following basic organisational parameters:-

Commitment
Cognisance
Communication
Competence

These tend to find expression and amplification via "motherhood" issues such as Policy, Leadership, Safety Culture, Training, Coaching, Communications, Self-analysis, Audit and so on. The reader is then left to assess whether the issue affects his particular organisation, and, if so, how he should decide how to introduce the necessary corrective actions.

Some clarity is starting to emerge from the profusion of research investigations (Commission of the European Communities, 2000), and the NII now has a generic model for Safety Management Systems, an insight into the relationship between Business Excellence and Safety Culture, an approach to assessing the impact of organisational structures and a developing method for assessing compliance at a task level. Work is also underway to investigate the impact of Peer Review on Nuclear Safety.

NEW TECHNOLOGY

In the UK, the principal technology challenges are those associated with soft systems. The general rate of innovation has slowed over the last decade, whilst overall dependence on computing systems has increased, with the result that "smart" sensors, distributed control systems and "intelligent" information management systems are playing a major role in determining how processes are controlled. Despite the greater reliance on such systems, the available evidence suggests that reliability has improved in a general sense although specific failures have highlighted the need for even greater attention to detail during design, procurement and commissioning. The impact of the "Millennium (Y2K) Bug" was anticipated well in advance and suitable remedial actions were taken where necessary. In the UK, only three failures were noted across the industry in the immediate post-Millennium period, none of which were in safety-critical systems and a similar number of non-safety-critical failures were observed following the 29 February 2000 transition (Leap Year Problem).

Assessment techniques have been sharpened up and emphasis has now shifted from a degree of reliance on the robustness of systems design assumptions to the role of the systems assessor in providing concrete evidence that systems will function to specification under all conceivable multiple fault conditions.

A key learning point to emerge from the Y2K experience has been the additional operational insight into just how reliant some systems can be on the experience and knowledge of those who are involved in design, procurement and commissioning processes. Thus, perhaps somewhat ironically, the principal technology challenge, at present, appears to be people-related rather than directly technology-related. This problem finds expression in the maintenance of corporate memory, the training of engineers and the development of analytical techniques. When coupled with the industry's desire to downsize, reduce its investment in new projects and refurbish systems at the lowest possible cost of ownership, these people factors probably represent the major technology challenge at present.

The effective maintenance of corporate memory and the comprehensive training of engineers are almost certainly the key technology challenges for the foreseeable future. A number of initiatives are underway to deal with these two issues. Adoption of Business Excellence principles by the key stakeholders is starting to have some impact because the importance of corporate memory is highlighted in the process. The "Investors in People" initiative across British industry is also having some impact on the general recruitment and development of engineers, but there is, nevertheless, a downward trend in nuclear training which may be expected to have an impact on prospects for the future (Furness, 1999b). The NII is currently working on a strategy to address the medium/long-term training problem and it has put in place interim remedial actions to ensure that continuity is maintained. As well being a logistic and resource problem in the short-term, however, this is also a motivational problem because long-term career prospects play a significant role in determining recruitment and human resource development. Another key activity, therefore, is to ensure that suitable career prospects are available for all, and this is being addressed via stakeholder personnel development and enhancement programmes.

Maintaining an equivalent peer capability in the NII is also a challenge, and is obviously influenced by the industrial climate and skills market in which the Inspectorate operates. The Inspectorate does not have the infrastructure to offer new graduates the necessary breadth of training to enable them to achieve Chartered status within their specialist technical disciplines. It has traditionally relied on recruitment of staff who are already of Chartered status, generally in their thirties or forties. Thus at a time when the UK nuclear industry has been subject to some criticism from the Inspectorate for having cut its staff too far and is seeking to recruit, the Inspectorate finds itself competing for the same scarce resources. Clearly the ability to recruit and retain the services of suitably qualified and experienced discipline specialists is driven by the market. Yet again, the technology challenge, this time for the Inspectorate, appears to be dominated by people-based issues.

STANDARDS, PROCESSES AND CONTINUOUS IMPROVEMENT

Mention has already been made of the continuous improvement processes that are being applied across the industry. These include Quality Assurance programmes, Training Needs Analysis, Business Excellence techniques and, where appropriate, Business Process Re-engineering.

The standards derive from many years of operating experience and address a wide range of design, assessment, inspection, operation and decommissioning activities. The licensees keep their standards and procedures under review, and NII inspects the arrangements for their production and servicing. For its part, NII has an equivalent range of standards, including Approved Codes of Practice, Guidance for Site Inspectors, Business Management Systems, Safety Assessment Principles and Technical Assessment Guides which are reviewed on a regular basis. In the human factors area, NII currently has under development guidance on topics such as the assessment and control of fatigue and the HSE produces guidance on topics ranging from display screen equipment usage through to personal protective equipment.

Extensive research is undertaken by the HSE, NII and the UK nuclear Industry Management Committee (IMC) (a consortium of nuclear power generators) so as to develop and refine techniques for continuous improvement. The IMC commissions research in response to safety issues identified by NII in a Nuclear Research Index. which is produced by the NII, on behalf of the Health and Safety Commission. As well as promoting research via the IMC, the NII also undertakes and commissions its own research via a Levy Programme. Recent research results from the Health and Safety Commission Coordinated Programme include guidance on the design of alarm handling systems, situation awareness, accident management, shiftwork, colour coding, incident analysis and human reliability assessment. Current research is focused on, amongst other things, violation control, downsizing, team management, personal protective equipment and understanding the relationship between occupational safety and nuclear plant safety.

On the process side, within the licence conditions, there is the requirement for a "safety case" for the installation. An important aim here is for the licensee to justify the safety of what they are doing. This allows the possibility that a wide variety of activities may be licensed, given an appropriate safety case for each. In the context of a changing organisation, the safety case also has the vital function of capturing for corporate memory the reasons for particular design features, operating rules and instructions, procedures, etc.. It also provides a clear benchmark for variations, modifications or separate stages in the life of a plant, so that changes can be properly analysed, designed and assessed. NII takes a particular interest in the licensee's own process for managing safety case changes, including the production of the case for change, review arrangements, formal endorsement by their Nuclear Safety Committee and ensuring that the licensee takes full ownership of the safety case, if they have not prepared it themselves (Pape, 1996).

Other important NII functions relevant to achieving continuous improvement in a changing industry include maintaining strategic oversight, overseeing and promoting the nuclear safety research programme, making input to policy development, contributing to international work and keeping the public informed about nuclear safety.

NEW FRONTIERS

Maintaining and improving nuclear safety is everyone's concern. The new frontiers for regulatory interaction include making certain that commercial pressures are compatible with, and promote, the overall safety objective, ensuring that suitably qualified and experienced persons are readily available at all times, ensuring that licensees are aware of the magnitude of the people-based safety problem and ensuring that suitable techniques are available on both sides of the regulatory fence to provide high confidence that the industry is regulated in an effective manner.

Because regulation based on detailed rules had been found by Robens to be unable to keep up with the pace of change in industry, the UK health and safety regulation system has been essentially risk-based since the Health and Safety at Work Act 1974. The HSWA introduced the basic principle that risks must be reduced so far as is reasonably practicable and accordingly it adopted an approach which set goals, based on risk reduction. Thus, from its inception the HSWA has been concerned with "risk-based regulation". This term has come to be used in the nuclear industry in a more specific sense, to imply a more prominent use of PSA and possibly Cost-Benefit Assessment. In the UK, the fundamental basis of good safety, as expounded in the Tolerability of Risk (HSE, 1992a) and the Safety Assessment Principles (HSE, 1992b) documents is sound engineering principles and management. PSA is then applied as a review technique to ensure that all foreseeable faults have been taken into account and, if necessary, protected against. Also, PSA can show that there is no undue reliance on a particular (uncertain) feature; and it can be used (with caution) to show that the overall risks are low. PSA facilitates the demonstration in the Safety Case that risks are As Low As is Reasonably Practicable (Pape, 1996).

Risk-based regulation provides a framework for making informed judgements. One of the new frontiers for making better use of this approach will be to ensure that safety cases better reflect the risk arguments and that plant and operational conditions are fully consistent at all times with the risk assessment assumptions. Some improvements are underway in this particular area and engineering substantiation and walkdown methodologies, for example, are being used to provide stronger linkage between risk assessment theory and practice. This is a new frontier for regulatory interaction because the approach is not yet uniform and because the method of inspection is qualitatively and quantitatively different to those which are used for normal compliance and assessment inspections. The approach requires specialist knowledge of multiple systems as well as extensive contact with process operators and designers to ensure that the full coverage is achieved. This is challenging for the regulator because of the resource overhead that the activity necessarily creates and it is challenging for the licensee because it requires access to parts of the organisation that might not normally engage in any extended interactions.

Another frontier concerns the matter of organisational change and the means by which this is to be monitored and controlled. For the licensee, there is an initial overhead associated with assessing its "baseline" resource requirements for safe operation and a continuing overhead associated with making and implementing adequate arrangements to control organisational structure and resources which might affect safety. There is also the problem of selecting suitable tools for such assessments and, although change procedures are in place, these need to be supplemented by more informative processes which provide early insight into potential problems. For the regulator, there are two basic problems. The first is concerned with ensuring, on a sample basis, that licensees are fully compliant with the provisions of the relevant Licence Condition and the second is concerned with ensuring that the change process fully reflects the safety implications at all stages. These are nontrivial problems for licensee and regulator alike.

There is also a frontier for regulatory interaction which is associated with the general management of the industry and, in particular, the effect that downsizing has had on personnel. Although some improvements in deployment have been noted, there have also been very significant changes in fatigue and stress. Excessive shiftworking has been in evidence and there have been cases of long hours working which have taken their toll on vulnerable individuals. Some licensees have been unable to operate without a long hours culture and they have also found it necessary to contractorise and outsource a number of essential functions which have occasionally produced ambiguity with respect to overall responsibility. This represents another challenge to both licensees and regulator.

For licensees, it will be necessary to monitor much more closely the impact that changed working practices are having on their employees and for the regulator it has been necessary to specify the indicators that will be used to track such phenomena. This, in turn, has necessitated the generation of suitable guidance, together with a greater appreciation of the significance of a wide range of work hygiene factors. The frontier that is represented by these respective roles is the need to create a mutual and informed understanding of the significance of human, organisational and management factors in a wide range of individuals as rapidly as possible so that interactions will be as effective as possible.

A final and vital frontier is the need to create a climate in which regulatory interactions are seen to be constructive and effective. This is important for licensees and regulator because, for the long-term, it is essential that satisfying careers can be perceived as being a realistic possibility in both camps. This is a particularly challenging frontier because the industry is under scrutiny and pressure to reduce costs and needs to be seen to be a worthwhile occupation. In order to maintain nuclear safety, it is essential that mutual respect is developed further and that suitable training is encouraged for the long-term (Furness, 1999b). For the licensee this requires a less adversarial perception of the respective roles, and for the regulator, this requires consistent, proportionate and transparent actions which are seen to be fair and reasonable. This frontier can only be crossed via increased investment in training and improving the levels of trust. However, as in any trust game, this only works when there is consistent delivery. The challenge to both licensees and regulator, therefore, is to find suitable ways to develop trust, whilst retaining the necessary independence of thought and action that is required for effective interaction and the maintenance and improvement of nuclear safety.

ACKNOWLEDGEMENTS

The author would like to thank his colleagues for their helpful comments during the preparation of this paper. It should be noted that the views expressed are those of the author and do not necessarily reflect the views of the Health and Safety Executive.

REFERENCES

Commission of the European Communities (2000). *Organisational Factors; Their definition and influence on nuclear safety.* Final Summary Report, Contract FI4S-CT98_0051, Brussels

Furness, B.J. (1999a). *Nuclear Regulation into the Millennium*, presentation to the NII 40th Anniversary Seminar, Queen Elizabeth II Conference Centre, Westminster, 24 November 1999.

Furness, B.J. (1999b). *Nuclear Regulatory Challenges or Who we should train and why - A Regulatory Perspective.* OECD Workshop on Assuring Nuclear Safety Competence into the 21st Century, Budapest, Hungary, 12-14 October 1999

Hollnagel, E. (1994). *Human Reliability Analysis: Context and Control.* London: Academic Press.

HSE (1992a). *The Tolerability of Risk from Nuclear Power Stations.* HMSO, 1992; ISBN 0 11 886368 1.

HSE (1992b). *Safety Assessment Principles for Nuclear Installations.* HMSO, 1992; ISBN 0 11 882043 5.

HSE (1994). *Nuclear Site Licences under the Nuclear Installations Act - Notes for Applicants.* HS(G)120. Sudbury, Suffolk: HSE Books.

HSE (1996). *Managing for Safety at Nuclear Installations.* Sudbury, Suffolk: HSE Books, 1996; ISBN 071761185X

Pape, R.P. (1996). *Nuclear Regulation in a Changing Environment.* Proceedings of BNES Congress, 4-5 December 1996.

Taylor, F.E. (2000). HMNII view of the Licensee as an Intelligent Customer. *Nuclear Energy*, **39**, 3,175-178).

Taylor, F.E. and Coatsworth, A.M. (1999). *Partnering in the Nuclear Industry*: A Regulatory Perspective. British Nuclear Industry Forum Conference on Alliancing, Glasgow, June 1999.

Taylor, F.E. and Turton, D (1998). Regulatory requirements for the use of contractors on nuclear licensed sites. *Nuclear Energy*, **37**, 1, 55-58.

Weightman, M. (1999). *Regulation of Organisational Change in the UK Nuclear Industry.* IAEA Technical Committee Meeting - Canada 1999 on Integrating the Management of Safety and Successful Business management of Nuclear Power Plants.

The next chapter, by Reiman and Norros, is a detailed analysis of the issues which confront the nuclear power industry regulator in Finland. The paper is distinctive in that it does not highlight prescription versus goal direction as the central dimension of regulatory variation. Rather, it sees the inspectorate as performing three distinct roles: the role of expert, providing information to the industry, the authority role, imposing regulatory requirements on the industry, and a third role of communicating to the public about the safety of nuclear facilities. The dilemmas for this regulator stem from the need to balance these three, sometimes competing roles. The paper reports on a study of the organisational culture of the regulatory inspectorate which investigates how the inspectorate achieves the required balance.

11

Regulatory Culture: Balancing the Different Demands of Regulatory Practice in the Nuclear Industry

Reiman, T. & Norros, L

Abstract

A case study to investigate the organisational culture of the regulatory authority was conducted at the Radiation and Nuclear Safety Authority of Finland's (STUK) Nuclear Reactor Regulation (YTO) – Department. Organisational culture is defined as a pattern of shared basic assumptions, which are basically unconscious. A combination of quantitative and qualitative methods was used in the research. In this paper we propose that the regulator has an indirect influence on the safety culture of the nuclear power plants. This indirect control is an important aspect of the regulatory culture. Based on the results of the case study, we propose a model of the regulatory culture, comprising of three occasionally conflicting roles: the authority role, the expert role and the public role. The implications of these roles and their conflicting demands are also discussed.

Nuclear Energy and its Regulation in Finland

Finland has four nuclear reactor units in commercial use. They generate 30% of electricity in Finland. The first unit (Loviisa nuclear power plant, VVER-440 type pressurised water reactor unit) started operation in 1977, and in the early 1980's all four units (two in Loviisa and two in Olkiluoto) were in use. Loviisa power plants are owned by Fortum Corporation and Olkiluoto power plants (ASEA-ATOM type boiling water reactor units) by Teollisuuden Voima Oy. At the time of the writing of this paper, Teollisuuden Voima Oy had just issued the license application for the fifth reactor unit.

Regulation of the use of nuclear energy in Finland is based on the Nuclear Energy Act (990/87) and Decree (161/1988). These give Parliament the final say on building new major nuclear installations, including waste disposal facilities. They also define the licensing procedure and the conditions for the use of nuclear energy, including waste management, as well as the responsibilities and authority of the Radiation and Nuclear Safety Authority of Finland (STUK). Regulation of the radiation practices is based on the Radiation Act (592/91). The Electricity Market Act (386/1995) opened access to distribution networks and allowed foreign ownership in electricity supply.

The ultimate quality objective of STUK is to keep the radiation exposure of the Finnish citizens as low as reasonably achievable (ALARA principle) and to prevent radiation and nuclear accidents with a very high certainty (Safety As High As Reasonably Achievable, or the SAHARA principle). According to STUK's quality policy, STUK is "a regulatory authority, research institution and expert organisation" (STUK 1999). STUK operates under the administrative control of Ministry of Social Affairs and Health. The operating organisations are responsible for the safety of nuclear power plants. In accordance with the inspection programme it has formulated, the department of Nuclear Reactor Regulation (YTO) verifies that the operations and related support activities of the power companies are appropriate and in compliance with safety requirements. STUK produces regulatory guides called YVL –guides. There are currently about 60 YVL guides in the following eight series: general guides, systems, pressure vessels, civil engineering, equipment and components, nuclear materials, radiation protection and radioactive waste management. The power companies must follow the rules set in the regulatory guides unless they can prove they can achieve the same level of safety with other methods. (STUK 1998, STUK 1999.)

Regulatory control of operating nuclear power plants contains reviews and inspections, which can be divided into three categories as follows:

- periodic inspections are specified and registered by STUK in a plant-specific programme,
- inspections which the power company is obliged to request in connection with measures carried out at the plant or which STUK conducts at its discretion, and
- safety assessment based on operating experience and on safety research as well as other information obtained after the granting of the operating licence.

(YVL 1.1, 27 Jan.1992.)

ORGANISATIONAL CULTURE AS A CONCEPT

In recent years, the scope of safety thinking has expanded from a concern with purely personal characteristics to the impact of the workplace and of organisational and managerial factors (Misumi et al. 1999, p. xxi). In the nuclear field, a term *safety culture* (IAEA 1991) has come into use to describe the attitudes and values required for a reliable and responsible worker. The attitudes emphasise questioning, rigorous and prudent approach and communication as a basis for sound safety culture. The same safety guide emphasises the regulator's role as an open and co-operative agency (IAEA 1991).

In the literature *organisational culture* is defined as "[a] pattern of shared assumptions that the group learned as it solved its problems of external adaptation and internal integration ... [and is] taught to new members as a correct way to perceive, think, and feel in relation to those problems" (Schein 1992, see figure 1).

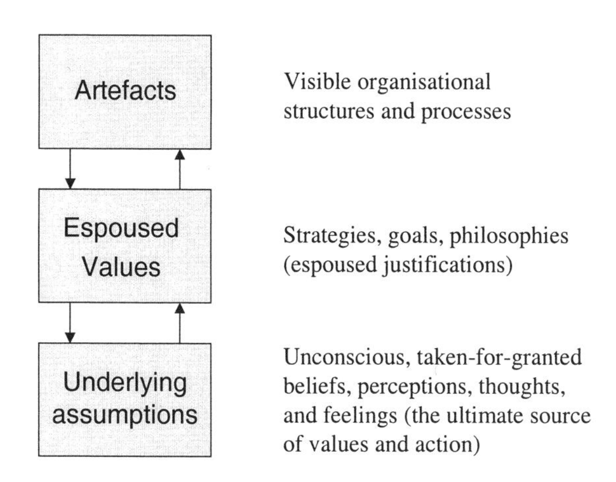

Figure 1. Schein's (1992) model of organisational culture.

According to Schein's model (see figure 1), the deepest layer of organisational culture consists of shared tacit assumptions that have resulted from a joint learning process. These assumptions make an individual's life predictable and meaningful in organisational context. The assumptions, which are basically sub- and preconscious values and beliefs, concern an organisation's strategy, goals, language, social interaction and leadership. The assumptions also consist of beliefs about the nature of knowledge and human nature and about the proper way to allocate rewards and status. Espoused values refer to conscious justifications to action. They predict what people will say in a variety of situations, but if they are not congruent with underlying assumptions, they do not necessarily predict what people will actually do in different situations. The surface level of culture consists of artifacts, which include the visible behaviour of the group and organisational processes, products and technology. (Schein 1985, 1992, 1999.)

Cameron and Quinn (1988, 1999) have proposed a theoretical model called competing values framework. According to the theory, organisations can be viewed along two dimensions (see figure 2):

- focus on internal processes versus focus on external processes
- focus on control versus focus on flexibility.

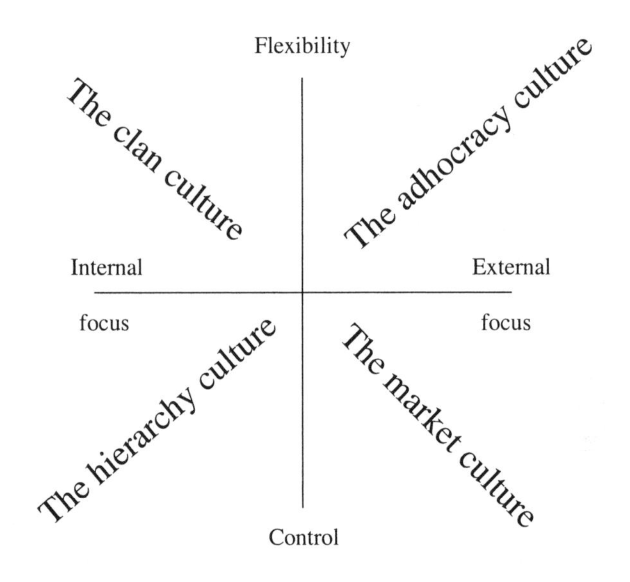

Figure 2. Cameron and Quinn's (1988, 1999) competing values framework

Together these two dimensions form four quadrants from which four dominant culture types emerge. Each dominant culture represents a distinct set of organisational effectiveness indicators.

They define what is seen as good, right, and appropriate. They represent the core values that guide action and decision making at the company. In a hierarchy-focused culture (see picture 2), procedures govern what people do and stability, predictability and efficiency are considered as long-term concerns of the organisation. The market culture values productivity and competitiveness by strongly emphasising external positioning and control. The workplace is highly results-oriented. The clan culture values cohesion, participativeness, teamwork and commitment. The adhocracy culture has as a major goal the fostering of adaptability, flexibility and creativity. Readiness for change is advocated. (Cameron & Quinn 1999.)

Regulatory culture is formed from an attempt to fulfil the requirements that the regulator's main duty sets and at the same time, guarantee internal cohesion and integration, as depicted in Schein's model (see figure 1). Regulatory culture is an important variable affecting the overall effectiveness of the regulation and the safety of the regulated system.

Organisational Culture at Nuclear Reactor Regulation (YTO) – Department

A case study to investigate the organisational culture of the regulatory authority was conducted at the STUK's Nuclear Reactor Regulation (YTO) – Department. Objectives of the study were to conceptualise and describe the main characteristics of YTO's organisational culture and to carry out a tentative core task analysis of the inspectors' work. A combination of quantitative and qualitative methods was used in the research. The first phase of the study consisted of a survey method (organisational culture survey and job motivation and stress survey) and the second phase consisted of a workshop for the whole staff. Document analysis and preliminary interviews were conducted in order to get some background information about the organisation. Interview questions were formed on the basis of the document analysis and literature review. Questions primarily addressed knowledge management related themes: How STUK gathers and disseminates information and from where does it receive feedback on its effectiveness. Also the questions addressed the core mission and core tasks of the department of Nuclear Reactor Regulation.

The results of the interviews and document analysis were used in tailoring the questionnaire (FOCUS, First Organizational Cultural Unified Search by De Witte & Van Muijen 1994) to better fit this kind of organisation. Some context-specific questions were also added. The questionnaire was based on the Cameron and Quinn's model (see figure 2). After modifications, the extended FOCUS-questionnaire was distributed to the whole staff of YTO. Response rate was 68 percent, which means 36 subjects. Data was factor analysed and summated scales were formed. In the second phase of the research, a safety culture and development workshop was carried out. The workshop was organised twice so that everybody could have a chance to attend one or the other.

Participation at the two workshops was over 90 percent altogether. Topics for the workshop were iterated from analysis of results of the preliminary investigation and the survey study. Topics were also discussed and developed with a planning group consisting of five members, three from STUK and two from VTT (the authors). In the workshop the following topics were addressed:

- recent and potential changes in the working environment
- special requirements of the regulatory work
- quality of work, definition of quality in regulatory work
- values, "what should be the core values which guide the decision making at STUK?"
- effects of regulatory decisions, guides and inspections on the immediate safety and long term safety culture of the Finnish NPPs
- actions needed to enhance STUK's own organisational culture and safety culture of Finnish NPPs.

Results of the Case Study from the Cultural Point of View

The questionnaire results clearly indicate YTO to be a hierarchy-focused culture (see figure 2) with less emphasis on innovation (the adhocracy culture) and results or goal setting (the market culture). Statistical tests (ANOVA) showed that the differences between the mean scores on the newly formed hierarchy variable and the other three variables (clan, adhocracy, market) were highly significant. The lesser emphasis on goal setting was felt by some respondents as increased ambiguity and uncertainty about his or her duties and how these contribute to the overall goals of the organisation. The results from the qualitative part of the research show that the employees at YTO value professional knowledge, openness, courage, fairness, efficiency, questioning attitude, teamwork and independence. These values can be interpreted as YTO's conception of an ideal regulatory culture (see figure 3). This ideal regulatory culture encompasses elements from all the dominant culture types of Cameron and Quinn's framework (see figures 2 and 3). When discussing targets for development the social aspects of work became central: social and professional support of co-workers, socialisation of newcomers (transfer of knowledge and experience), internal communication and internalisation of YTO's values (the clan culture in Cameron and Quinn's model, see figure 2). Resource and work-process management were also mentioned as targets for development. Requests for the clarification of the YTO's goals and the enhancement of feedback at both individual and organisational level were emphasised in both the survey and workshop results. The focus on hierarchy and rules was an expected result since bureaucracy and rules are the essence of democratic government . Furthermore, the safety critical nature of the work requires firm methods to assure the quality of the results.

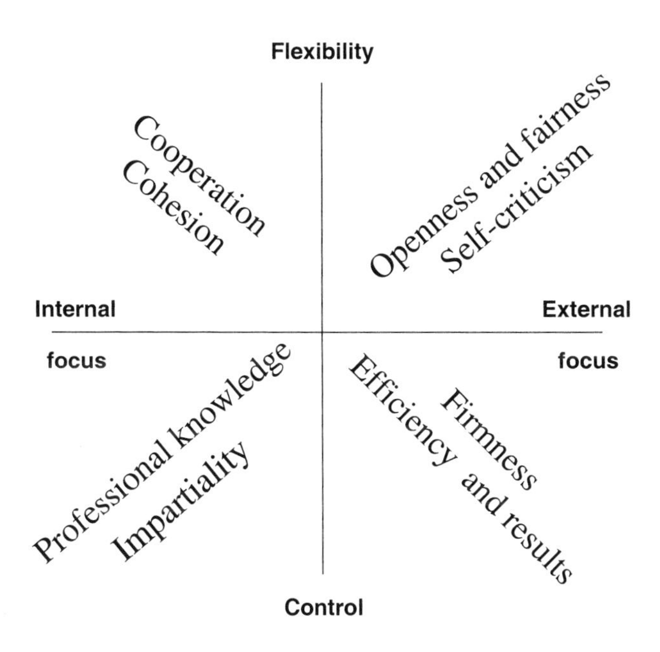

Figure 3. YTO's conception of an ideal regulatory culture

We can see from figure 3 that YTO's conception of an ideal regulator includes impartiality (regulator is not pro- or antinuclear and stands apart from exercising influence in energy policy matters). This requires openness both to the public and to the power plants. Openness means also accountability to the public. STUK has to explain and justify all of its actions and decisions to the public and to the power companies. At the same time, the regulator has to be firm (but fair) and achieve results by being efficient and by having a high level of professional knowledge. At the individual level, this knowledge acquisition, sharing and dissemination require self-criticism and humility concerning own level of knowledge. The sharing and dissemination of knowledge also requires co-operation and a high level of cohesion among the personnel (shared values). The results of a study ordered by STUK show that the power companies perceive STUK as open and flexible (Sinkkonen 1998). According to this study, self-criticism (which also means innovative thinking), efficiency and results on the external dimension and co-operation and cohesion on the internal dimension require more attention. Efficiency does not mean maximising the net return of capital on the organisation, but timely (fast) handling of documents requiring YTO's approval or review. Placing value on results means that the guidelines and decisions made at YTO should (visibly) enhance the safety of nuclear production.

In this study, the biggest threat to YTO's organisation was conceived to be excessive bureaucratisation (see also Quinn 1988). Bureaucratisation means the formation of routines that are taken-for-granted and so are not questioned even if they are no longer appropriate. Also the meaning of one's work is easily lost in a bureaucracy. Bureaucratisation and routinization is typical of organisations that operate in highly stable and predictable environments (Morgan 1997). Bureaucratisation can also cause the working environment to look like it was stable when in fact only the routines of the organisation are stable. This can increase an organisation's resistance to change, because no need for change is perceived among the personnel. Also, because it operates in the public sector, there are no market pressures forcing YTO to change. The work environment of YTO has been very stable, but recently the deregulation of the electricity market, ageing of technology and personnel and modernisation and introduction of new technology (programmable automation for example) have brought new uncertainties. At the workshop, when discussing recent and potential changes in the work environment, the biggest changes perceived were globalisation, technological changes (ageing, new technology), social changes (attitudes), and changes in nuclear policy (deregulation, future of nuclear power).

Another threat to YTO's culture that was identified in this study is a lack of direct feedback both from internal and external sources. This can be seen from the targets for development (from the workshop), which emphasise the clarification of YTO's goals and the enhancement of feedback at both individual and organisational level and internal communication. The implications of this are discussed in the following sections.

DEMANDS ON THE REGULATORY PRACTICE

The above-described results were achieved under the assumption that culture has a double role of maintaining external adaptation and internal integration. The actions in the organization reflect these aims and they can be defined through a model composed of two dimensions along which actions in the organization may vary. The competing values framework of Cameron and Quinn define the internal-external dimension to characterise the typical focus of actions, and the control-flexibility dimension to characterise the type of control the actions manifest. In this framework the YTO department is clearly a rule- and hierarchy-oriented organization that emphasises control and is focusing on internal processes. The results indicated, however, that within the organization there might exist subcultures, which can be interpreted as expressions of different ways of coping with the uncertainties in work. Analysis of the material produced by the personnel in the development workshop gave further information about the culture. It was found out that the inspectors feel that they receive only very little feedback on their actions and that their practice is based on individual expertise. These conceptions of the task demands explained the strong emphasis on internal focus in the organization in the results . We learned also that owing to changes in the external environment the personnel experience a need for greater flexibility in regard with both internal and external actions. When interpreted within the Cameron and Quinn's framework, a more elaborate description of the organizational culture could be achieved.

The results of the development workshop were also interpreted in a further framework. Instead of considering YTO as an organization in general, it was addressed as a *particular activity system* (Engeström 1999). Accordingly, we were interested in the tool-mediated person-environment interactions that the inspectors conduct in their daily work and the meanings they attach to these interactions. The system is understood as a unity and its activity is seen from the ecological perspective as promoting the survival of the system (Järvilehto 1998).

Knowledge of the activity system can be achieved through analysis of the *practice of coping with the demands* of this particular work. From this perspective the cultural regularities described with the help of the Cameron & Quinn's model are not laws according to which people in the organization act. Instead they are results that people produce when acting within the organization. They are facts that persons can take into account according to various logics (Eskola 1999). These logics refer to different value frames that give personal meaning to the objects of environment. Understanding practice means identifying these "choices" as they are made in daily work.. The results concerning the YTO-department could thus be interpreted as inspectors' different ways of conceptualising and coping with the demands of nuclear reactor regulation.

The core task analysis

The contextual features of work that actors may take into account can be analysed historically and currently. The core task analysis is a method that we have developed for the analysis of current practice. (Norros & Klemola 1999, Norros & Nuutinen submitted).
The core task analysis means modelling the content of the tasks in regard with two aspects. These are the functional aspect, and the intentional aspect that has to be defined in a situated way. The method of analysis has connections with the functional modelling concept of Rasmussen (1996) that he has applied to describe complex human-environment systems and their control demands. In our analysis the human-environment system is conceptualised from the perspective of the cultural-historical theory of activity (Leontjev 1981) as an activity system (Engeström 1999). In the core task analysis we first define the object of activity in terms of the results aimed for. In respect of the functional aspect we then define the critical functions of the object and the boundary conditions for appropriate control of the system. At the same time we elaborate critical roles of the agents. In

respect of the intentional aspect of activity, situationally specific goals and tasks are derived from the object. Because all aspects of the task are not known a priori it is necessary to conceptualise the task from the point of view of actual practice. We therefore mirror the critical functions and boundary conditions as possibilities and constraints that set psychological demands on action. Action itself is viewed as the agent's tool-mediated interaction with the environment. The observed performances can be interpreted as habits of action in a semiotic analysis of the reasons for actions. In this case we limited ourselves to the conceptual analysis of the tasks of the inspector of the nuclear reactor regulation department (YTO). The basic elements of the core task are developed in the following and the model presented in the last section is elaborated through the results of the organisational culture of YTO.

The critical functions and the roles of inspectors

The activity of the nuclear reactor regulation department is focused on the use of nuclear energy. The objective is to secure a safe use of nuclear energy. There are three critical functions that are central in fulfilling this objective: YTO has been given a *mandate from the society*, to *use authority over the operating plants* in such a way that *nuclear energy is used according to law.*

We now elaborate these critical functions and analyse the roles that are connected with fulfilling of each function. We start from the last function because it relates directly with the use of nuclear energy.

The function of ensuring that nuclear energy is used according to law implies an *expert role* for the inspectorate.
The various tasks of the inspectors that are carried out to fulfil objectives of the YTO all encompass the demand on expertise in different fields of nuclear energy, the central one being the setting of safety regulations and verifying the compliance with them through inspections. The expert must show good judgement in his practice. The inspector is expected to posses a high level of knowledge in his special domain, and he is personally responsible for evaluating his competence and updating it according to the precision and increase of knowledge in his domain. This is especially problematic because judgements involve social as well as technical aspects. It is important to note that the ALARA and SAHARA principles, mentioned earlier, ultimately depend on societal judgements about what is "reasonably achievable".

The inspector's work is a classical example of the psychological demands of professional judgement. From the point of view of the epistemology of practice, in contrast to the model of technical rationality explained by Schön (1988), expertise requires methodically controlled thinking and an epistemic interest towards the object of action. As a result knowledge of the object is constructed in the practice. Connected with the practical habits of constructing knowledge, an awareness of one's personal conception of knowledge should also develop.

For construction of knowledge it is important to respect the object through establishing a dialogical relationship to it (Megil 1997, Klemola & Norros 2000). The object should have the possibility to be heard, not limited by pre-conceptions. This requires consciousness of the effect of one's own concepts on the perception of the object (Dewey 1929, Boudieu & Vaquant 1992). It is equally important to accept practice as knowledge and to avoid total rejection of intuitive knowledge due to difficulties in justifying it (e.g. Polanyi 1964). The expertise of the inspector is a result of aiming at the best knowledge possible with the help of experimental thinking (Dewey 1933). Certainty as such should not be the target. The quest for certainty may turn out to be disadvantageous, if it leads to lack of self-confidence and search for methods to cope primarily with that problem.

A realistic attitude to knowledge also encompasses the idea of dialogue and co-operation. This is especially necessary in the context of nuclear reactor regulation. The expert is also assumed to be in control of his own resources and able to act under pressure, which become an issue in critical situations where there is a need for prioritising tasks, sharing attention, and managing time.

The second critical function of the task relates to the use of authority over the operating plants. The bases of the regulatory actions are defined in the acts and decrees and international agreements. In its quality manual STUK (1999) has defined a number of principles that should guide regulatory practice. These are legitimacy, openness, independence, equality, sense of proportion, verifiability, and boundedness by intention. These principles can be interpreted as defining the second role of the inspector, the *authority role*. The major task is control, and it is realised through specifying safety regulations and verifying the compliance with them through inspections.

The authority role sets essential psychological demands on the inspectors' practice. The operating plants are responsible for the safe use of nuclear energy independent of the extent and manner of regulatory control. This distinction between responsibility for safety and safety control creates a significant demand on the control practice. The inspectors must develop different means like indicators, sampling schemas, inspection schedules etc. for indirect observations of the plant operation. Actions are indirect control measures, and feedback from their own actions is mediated and often very scarce.

The development of the art of mediated control is a challenge for the inspector. It also requires mastery of using rules and regulations as means of control. Owing to the fact that rules are general descriptions they never fully capture the particular situation. Consequently, they cannot be applied directly but used as means to make sense of the situation and to act accordingly. The inspector always makes judgements regarding the compliance with regulations, and he must also perceive things beyond the framework of the regulations. This assumes construction of knowledge in practice. The ability of the expert to do this develops in the authority role, which is also the professional practice of the expert. Reflexivity is a general quality of this practice. Essentially there is awareness of the societal boundedness of concepts and schemas of perception (Bourdieu & Waquant 1992) and of the interaction between the concepts and the object (Dewey 1929). There is particular demand on a reflexive inspector in high reliability organisations in which sensitivity for safety relevant cues that are not yet clearly identified is essential. Through reflective practice the inspectors could perhaps operate within the incubation period (see Turner and Pidgeon 1997) and prevent the development of accidents.

The mastery of the principle of independence requires that the inspector is able to distinguish the factual content of an opinion from the subject who holds it. Actions should be based on the factual arguments. This is another aspect of reflexivity in regulatory practice.

The third critical function is fulfilling the mandate from the society. YTO is accountable for the control of nuclear safety to different sectors of society. The mandate requires that the operations of YTO are public and that relevant knowledge of the use of nuclear energy is communicated according to the needs of the citizens. In its communication and operations YTO is at the same time obliged to maintain secrecy in regard to confidential information on individuals, business secrets, security and similar matters.

The above mentioned demands bring up the third role of the inspector, the *public role*. Communication is the central task of this role. It is the responsibility of each individual inspector to inform the public of matters in his domain of expertise but communication also forms a separate function and department in the organisation. According to the set quality criteria information should be reliable, true and sufficient, offered in right context, and presented in generally comprehensible form.

The public role sets particular demands on practice. They are connected with the ability of the inspector to make judgements about what information he is obliged to communicate and how this obligation is restricted by the demand for maintaining confidentiality. Choices of the content of communication should be as independent as possible of the inspectors own attitudes towards the use of nuclear power. Communication also shapes the public opinion about nuclear energy but should not take the form of indoctrination. Thus, the mastery of the psychological demands set by the public role assumes not only independent situational judgement but also personal strength and courage. Mastery of these demands mean that sense of proportion is maintained. It depends on successfully balancing fairness and firmness in relationships with the operating plants on the one hand and openness and confidentiality in relationships with the public on the other hand. As a result the inspectors and YTO achieves credibility in the eyes of different interest groups.

According to sociological analyses of the modern risk society this task is not easy owing to the high level of awareness of risk by the public and lack of trust in the experts who might be relied upon for protection (Beck 1986, Giddens 1991).

As a summary of this analysis of the regulatory roles and the psychological demands they impose on the inspector, we identified three types of developmental tensions that need balancing in action. These are:
1) Developing personal insight of the core demands of the task as action-orienting principles
2) Balancing between the demands of each three roles
3) Balancing between stability and change.
The content of these developmental tensions will be elaborated in the next, concluding section.

ORGANISING FOR EFFECTIVE REGULATORY PRACTICE

The environment where YTO operates has been very stable, but now, deregulation, introduction of new technology and ageing of plants and personnel have brought new uncertainties. Transfer of knowledge and experience to newcomers (socialisation) was perceived as a big challenge. This means the articulation (see Nonaka & Takeuchi 1995) of tacit routines, assumptions and ways of working that have been formed as the organisational culture was being learned. Also the acquisition of new expertise and upkeep of existing expertise have become a challenge. These changes in the working environment set further development tensions for the three roles identified in the previous section. In figure 4 we have integrated the results of the two approaches used in this research. The figure depicts the core task of the nuclear inspector. It contains the three roles, the demands of the different roles and the interaction between the roles. The figure broadens our understanding of the ideal values and the dynamics between external versus internal focus and flexibility versus control. From the perspective of the core task, we can elaborate the model of regulatory culture. The role of expert demands flexibility and creativity (creation of new knowledge, see figure 2) and the role of authority demands control, measurement, documentation and processing of information. Professional knowledge is a highly emphasised value in YTO's culture, but its practical implications are not so clear. The regulator needs ways to maintain, develop, transfer and capitalise its expertise. The transfer between roles is also experienced as stressful among the department's personnel.

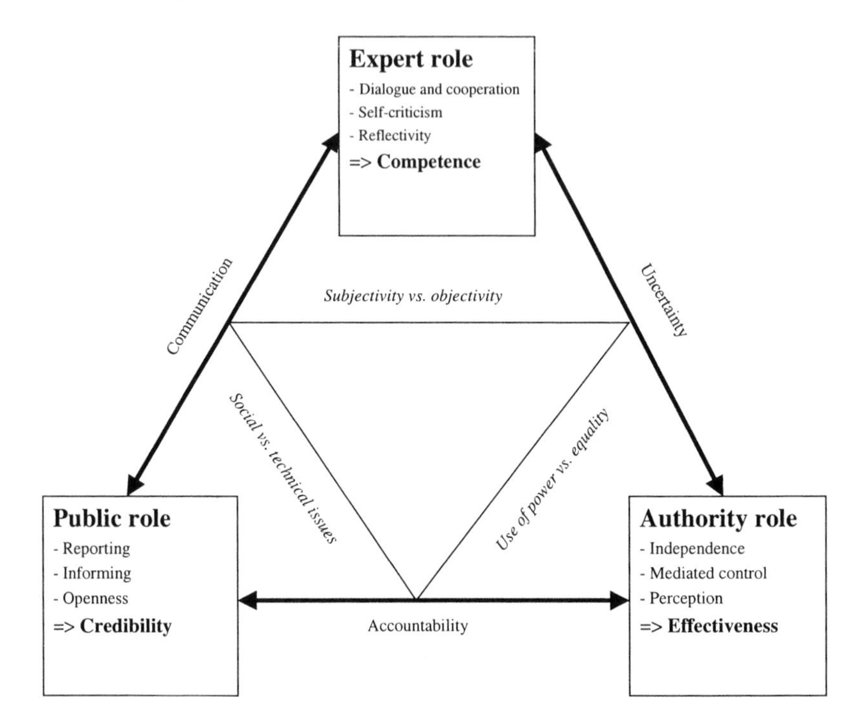

Figure 4. The core task of the inspector expressed as balancing the different roles and tasks

The three different roles and their demands are depicted in figure 4. Also the connections between the roles are presented. Fulfilling the requirements of public role leads to credibility, fulfilling the requirements of the expert role leads to competence and fulfilling the requirements of authority role leads to effectiveness. In a good regulatory culture, all the three roles must be in balance and appropriate support structures (management, human resources etc.) must be functioning. The reputation that regulatory organisations must maintain to retain public trust requires them to perform to the highest standards (IAEA 1999). This means that the requirements of all the three roles must be taken into account. The balancing between the roles is complicated by the inherent discrepancies between the roles. The expert role differs from both other roles in *subjectivity.* Knowledge is always subjective, but the regulator speaks (to the public and to the power companies) with the voice of objective authority. The communication required in the public role is different from that required in the expert role. *Uncertainty* is allowed in the expert role but not in the authority role. The authority role is the only role in which the *use of power* is allowed, and the public role demands taking into account *social issues* in addition to technical issues. *Accountability* sets a tension between the public role and the authority role, because the public role demands openness to the general public but the regulator has responsibilities also towards the power companies and cannot share all the information it has.

The balancing between the roles has implications also on the organisational level. The regulator has an indirect influence on the safety (and safety culture) of the power plants. Because the influence is indirect, it is hard to see the results of one's work. This means that the feedback from one's actions is also indirect. External feedback is often mainly negative. But at the same time employees at YTO clearly acknowledge the importance of their job. This can cause stress and lead to burnout. The results point out the importance of internal feedback from managers and colleagues. Feedback coming for the expert role is mediated through the authority role. The other expert at a power company easily sees the inspector as a representative of authority, not as an equal expert. Lack of external feedback sets even higher demands for a "mechanism" of knowledge creation and knowledge management (expert role). The indirect influence makes it hard to measure effectiveness and to define best practices, which further complicates the demands of authority role. The behaviour of the power plants acts also as an intervening variable in measuring effectiveness; YTO cannot easily differentiate their accomplishment from the power plants own development activities. YTO does not control all the variables that affect the safety of the NPPs. An important matter to address is how does the regulator put scientific research findings into use in regulatory practice and in making rules and regulations. Also, the role of (and interaction with) external research institutes and the regulatory authority's own research activities should be addressed. Feedback coming through the public and authority roles should be transferred and utilised also in the expert role, and vice versa. This is not a simple task. All the interest groups confront a representative of YTO not in one role at a time, but always in all the three roles, in varying degrees. For smooth co-operation, a certain amount of trust and respect should exist between YTO and its interested parties.

The learning mechanism of the regulatory agency is similar to the high reliability organisations' learning mechanism in that the regulator cannot operate by using trial-and-error learning (see Weick 1987). In contrast to typical errors in high-risk industries (Perrow 1984), errors and wrong decisions made by the regulatory body usually propagate slowly and cannot be easily traced to their origins, if such exists. This sets some unique requirements for development and maintenance of expertise in the regulatory authority (see figure 4). Also, a public sector agency may not accept critical external feedback, even when appropriate.. Wilpert et al. (1999) state this attitude provocatively: "People learn from mistakes, public servants do not make mistakes" (Wilpert et al. 1999, p. 136). The public role demands openness and consideration of social issues in addition to technical issues. The challenge for the regulator is to make others act in the appropriate way and to enhance the sense of responsibility of the regulated organisation. This is the only way to guarantee an adequate safety culture in the long run. Short-term safety may sometimes require more drastic interventions from the authority, but the goals should be focused upon the encouragement of a good safety culture at the regulated nuclear power plants. Long-term effectiveness demands fulfilment of the requirements of all the three roles.

The demands depicted in figure 4 could be used in developing training courses and qualification criteria for recruitment of new personnel. Also the demands and traits of YTO's culture that were conceptualised in this research can be used in planning further development projects at YTO. Additionally, the model of conflicting demands could serve as a basis for further research at other regulatory authorities.

REFERENCES

Beck, U. (1986). *The Risk Society: Towards a new modernity*. London: Sage.

Bourdieau, P. & Waquant, L.J.D. (1992). *An Invitation to Reflexive Sociology.* Cambridge: Polity Press.

Cameron, K.S. & Quinn, R.E. (1988). Organizational Paradox and Transformation. In Quinn, R.E. & Cameron, K.S. (ed.), *Paradox and Transformation. Toward a Theory of Change in Organization and Management*. Massachusetts: Ballinger.

Cameron, K.S. & Quinn, R.E. (1999). *Diagnosing and Changing Organisational Culture: Based on the Competing Values Framework.* Massachusetts: Addison-Wesley.

De Witte, K., Van Muijen, J. (1994). Organizational Climate and Culture in Europe. A Theoretical and Practical Introduction to the Focus Questionnaire (unpublished working paper). K.U.Leuven & V.U. Amsterdam.

Dewey, J. (1929/1999). *The Quest for Certainty. A Study of the Relation of Knowledge and Action.* Finnish translation. Helsinki: Gaudeamus.

Dewey, J. (1933). *How We Think.* Boston: D.C. Heath.

Engeström, Y. (1999). Activity theory and individual and social transformation. In: Engeström, Y., Miettinen,R. & Punamäki, R-L. (eds). *Perspectives in Activity Theory.* Cambridge: Cambridge University Press, 19-38.

Eskola,A. (1999). Laws, logics, and human activity. In: Engeström, Y., Miettinen, R. & Punamäki, R-L. (eds). *Perspectives in Activity Theory.* Cambridge: Cambridge University Press, 107-114.

Giddens, A. (1991). *Modernity and self-identity.* Cambridge: Polity.

IAEA, Safety Series No. 75-INSAG-4. (1991). Safety Culture. Vienna: International Atomic Energy Agency.

IAEA, TECDOC-1090. (1999). Quality Assurance Within Regulatory Bodies. Vienna: International Atomic Energy Agency.

Järvilehto, T. (1998). The theory of the organism-environment system (I). Description of the theory. *Integrative Physiological and Behavioral Science*, **33**, 4, 321-334.

Leontjev, A.N. (1981). *Problems of the development of the mind.* Moscow: Progress.
Klemola, U-M. & Norros, L. (2000). Logics of anaesthetic practice - outline of interdisciplinary methodology for studying the anaesthesiologists' activity in real life setting. Paper held at the 5th Naturalistic Decision Making conference, May 2000, Stockholm.
Megil, A. (ed.), (1997). *Rethinking Objectivity.* Durham: Duke University Press.
Misumi, J., Wilpert, B. & Miller, R. (eds.), (1999). *Nuclear Safety: A Human Factors Perspective.* London: Taylor & Francis.
Morgan, G. (1997). *Images of Organization* (2nd ed.). Thousand Oaks, CA: Sage Publications.
Nonaka, I. & Takeuchi, H. (1995). *The Knowledge-Creating Company: How Japanese Companies Create the Dynamics of Innovation.* Oxford: Oxford University Press.
Norros, L. & Klemola, U.-M. (1999). Methodological considerations in analysing anaesthetists' habits of action in clinical situations. *Ergonomics,* **42**,11,1521-1530.
Norros, L. & Nuutinen, M. (submitted). Studying process control as knowledge intensive work. European Work Process Knowledge -network final report.
Perrow, C. (1984). *Normal Accidents: Living With High-Risk Technologies.* New York: Basic Books.
Polanyi, M. (1964). *Personal Knowledge. Towards a post-critical philosophy.* New York: Harper & Row.
Quinn, R.E. (1988). *Beyond Rational Management.* San Francisco: Jossey-Bass.
Schein, E. H. (1985). *Organizational Culture and Leadership.* San Francisco.
Schein, E. H. (1992). *Organizational Culture and Leadership* (2nd ed.). San Francisco: Jossey-Bass.
Schein, E.H. (1999). *The Corporate Culture Survival Guide: Sense and Nonsense about Culture Change.* San Francisco: Jossey-Bass.
Schön, D.A. (1988). From technical rationality to reflection-in-action. In: Dowie, J. & Elstein, A. (eds.) *Professional Judgment. A reader in clinical decision making.* New York: Cambridge University Press.
Sinkkonen, S. (1998). Ydinturvallisuusvalvonta Imatran Voima OY:n ja Teollisuuden Voima OY:n edustajien silmin (in Finnish). [Nuclear safety regulation according to representatives of Imatran Voima Ltd. and Teollisuuden Voima Ltd.]. Master's thesis. University of Helsinki, department of social psychology.
Stewart, J. & Ranson, S. (1994). Management in the Public Domain. In Mckevitt, D. & Lawton, A. (ed.), *Public sector management. Theory, Critique & Practice.*
STUK. (1998). *Annual Report 1998.* Helsinki: Radiation and Nuclear Safety Authority.
STUK. (1999). *Quality policy.* Helsinki: Radiation and Nuclear Safety Authority.
Turner, B.A. & Pidgeon, N.F. *Man-made Disasters.* (2nd edition). Oxford: Butterworth & Heinemann.
Von Wright, G.H. (1998). *In the Shadow of Decart. Essays in the Philosophy of Mind.* Dordrecht: Kluwer Academic Publisher.

Weick, K. E. (1987). Organizational Culture as a Source of High Reliability. *California Management Review*, **2**, 133-127.

Wilpert, B., Fahlbruch, B., Miller, R., Baggen, R. & Gans, A. (1999). Inter-organisational Development in the German Nuclear Safety System. In Misumi, J., Wilpert, B. & Miller, R. (ed.), *Nuclear Safety: A Human Factors Perspective*. London: Taylor & Francis.

Yin, R. (1989). *Case Study Research: Design and Methods*. Newbury Park, California: Sage.

YVL 1.1. (1992). Finnish Centre for Radiation and Nuclear Safety as the regulatory authority for the use of nuclear energy. Helsinki: Radiation and Nuclear Safety Authority.

PART II: NEW FRONTIERS IN REGULATION

This section is entitled "new frontiers in regulation" in part because it contains two papers about areas of scientific/industrial activity where regulation is in its infancy (de Mol and van Gaalen; Baram).

Medicine is a very old ‘industry’ which has long been regulated by its professionals in a system of well-developed peer review. The breakdown of the trust in this system of self-regulation has led to the search for new paradigms. This has been particularly evident in the area of medical devices, as more and more equipment is implanted in, or attached closely to the human body. New technology combines with shifting societal values to create a new challenge to the regulator.

Biotechnology is a brand new area, which has catapulted the old industry of selective breeding into a new age, with many unknown hazards at the limits of scientific knowledge. It is also a rapidly changing area with huge potential profits at stake. Regulation has never been good at handling such combinations in the past and Baram’s chapter indicates that this is still a challenge which is not being met with success.

Finally we look at certification in detail. This has been increasingly seen in the past decade as a new tool for regulation and its use has been greatly expanded. Gundlach assesses the potential and limits for the tool.

The paper by de Mol and van Gaalen describes the advances being made in medical device technology. It demonstrates that the developments are being driven by commercial considerations and that patient safety has been a somewhat secondary consideration. Regulation is therefore vital. The authors identify three levels of regulation. First, legislation, second, activities by medical professional organisations and third, the interactions between doctor and patient. They discuss what is needed at each level to make it effective.

What is of particular interest in the present context is their description of the way in which a medical professional organisation, a professional safety board, has developed in the Netherlands to fill a void at the intermediate regulatory level. This is a form of industry self-regulation. De Mol and van Gaalen describe a safety management system which the board has developed and they argue that it has worked tolerably well. However they suggest that it might be preferable if these arrangements were prescribed by legislation rather than leaving them to the profession.

De Mol and van Gaalan essentially argue that safety in this area largely depends on the collection and dissemination relevant information abut risks and failure rates. Thus their recommendations at all three levels concern improved flows of information.

What lies behind this analysis is the idea that improved safety in relation to medical device technology depends on learning from mistakes and failures. There is an interesting difference here with many of the other industries described in this book. Most of these other industries are concerned with the prevention of major accidents which, because they are so rare, present few opportunities for organisational learning. Organisational learning is such contexts depends on identifying other data which can serve as indicators of how safely a system is functioning. In other words the prevention of major accidents depends on proactive safety management, relying on feedforward mechanisms. In the frontier area described by de Mol and van Gaalan, patient safety seems to depend more on developing reactive safety management which relies on improving feedback mechanisms.

12

MEDICAL DEVICE TECHNOLOGY AND PATIENT PROTECTION: CHALLENGES FOR REGULATION AND LEGISLATION

B.A.J.M. de Mol and G.L. van Gaalen

INTRODUCTION

Health care has always tried to adopt the most advanced technologies. This has led to sophisticated technologies for imaging, intervention, and life support, on which health care nowadays much depends. Prescription and application of these technologies are the prerogative of the doctor. Today, however, the developments in medical technology and the subsequent introduction of entirely new devices, professions, and services are on such a large scale, that regulation and legislation tend to lag behind more and more. As a result, many of the current much praised "improvements" in health care often carry the risk of increased rather than reduced patient vulnerability.

For many years, the relationship between patient and doctor could be characterised as a therapeutic alliance based on mutual trust (Gutheil et al, 1984). Harmful adverse events have always been possible though, as shown by studies on medical malpractice and negligence (Leape et al, 1991).

Traditional regulation and legislation on health care used to focus mainly on the duties of the state in terms of providing health care and organising it. The latter concerns matters such as medical training and certification, professional medical standards, and professional conduct statutes for application by medical courts.

Developments that have challenged the adequacy of this regulatory and legal framework are:

1) the change in relationship between patient and doctor into one with the emphasis on patient rights and patient protection;
2) the involvement in health care of other parties than doctors, in particular health care insurers, health care authorities, and medical device manufacturers. Law has meanwhile regulated the accountability of these non-medically trained individuals and they can thus be regarded as health care professionals, with the legal duty to fully develop their potentials and skills in the interest of patient safety.

In this chapter, we focus on the intertwining of medical device technology and patient protection and the challenges it poses for regulation and legislation. We hypothesise that regulation and control may take place at three levels: state legislation, rulings and standard setting by professional bodies, and safe behaviour and decision making by health care professionals.

HOW SAFETY FOCUSSED IS MEDICAL DEVICE TECHNOLOGY?

Health care owes its current improvements largely to progress in medical device technology, such as the development of life saving devices, like heart-lung machines and devices for haemodialysis (even disposable ones), and the development of sophisticated implants, like pacemakers, hip prostheses, and ocular lenses. Although the implants may seem somewhat less glamorous than live saving devices, they are generally regarded as indispensable to today's ageing population at large and they can thus be considered of equal importance.

Other examples are implantable drug delivery devices for the treatment of cancer or diabetes, which are becoming increasingly small and easier to fill for longer periods of time, and stents for the treatment of blocked coronary arteries, which are also becoming increasingly user friendly.

In order to tip the balance firmly into the direction of the benefits instead of the risks, research is done continuously. However, there will always be side-effects that are unpleasant and difficult to bypass, such as the need for re-implantation due to the limited durability of certain implants, and the need for life-long anticoagulation therapy after implantation of a mechanical heart valve. Side effects may also be harmful or even lethal. Pacemakers, for example, may halt when an anti-theft detection portal is passed, and mechanical heart valves may induce excessive blood clot formation, possibly leading to a cerebral embolism and subsequent death.

Some of the potentially harmful or lethal side-effects will soon be bypassed by technical advances:

- heart failure will be treated by means of a small, temporarily inserted pump,
- damaged heart muscle will be treated by means of injections containing the patient's own cardiomyocytes, and
- porcine heart valves will be denatured in order to serve as a scaffold for the patient's own stem cells for the creation of a quasi-natural valve.

Others, however, are harder to tackle. Xenotransplantation of hearts, kidneys, and livers of immunologically neutral pigs, for example, has virtually been stopped because of possible transfer of pig viruses onto humans. The public's awareness of a prion involved in mad-cows' disease has led to questions about the safety of using bovine material in heart surgery.

Despite all this, the push for rapid application of sophisticated new medical technology remains as strong as ever. The growing of cells for tissue repair, for example, is already being put into practice, even though it is still unknown how it can be controlled. We did not apparently learn much from a similar situation some time ago, when, after years of pleading by doctors, patients, and the media, a new treatment for Parkinson's disease was introduced, using embryonic brain cells instead of drugs. Although implantation of these cells led to the desired production of dopamine, efforts to control the process failed (Kohn et al, 2000), leading to dopamine overproduction and subsequent severe patient damage.

The above illustrates that there is an area of tension between the objective of treating patients with new technology and the price one may have to pay due to insufficient knowledge about the possible outcome. This is further complicated by the fact that today's development of new technology quite often automatically includes the involvement of venture capitalists. Each of them has their own, specific interests, and will do everything to be first, so that they can capitalise on the medical progress they seemingly made. So, the medical device industry often behaves like any other commercial enterprise, and patient safety is certainly not automatically its top priority.
The answer to the question "How safety focussed is medical device technology?" is thus , not very.

HOW CAN A COMPLEX HEALTH CARE SYSTEM BE CONTROLLED?

The medical device industry is a multi-billion dollar business, usually responding swiftly to the needs for a better and longer life for those who can afford it. It benefits particularly from welfare systems, which are supervised by governments or insurance companies or both, and which guarantee the availability of medical devices to millions of potential consumers. Even so, one cannot speak of a consumer market in the true sense of the word, because not the patient but doctors and other health care professionals negotiate about product quality, performance, safety, and costs. Patient power has thus been delegated. Even so, it remains the patient who has to pay the price of safety failures. The financial cost be compensated for by insurance, but not the human cost.

The level of awareness of the risk of personal harm differs among citizens from country to country. In the US, for example, there is a vocal patient safety movement, but not so in most European countries, although medical knowledge and care are much the same (Brewin, 1994).

Judging by the fact that the statements "first, do not harm" and "in dubio, abstine" are heard often in medicine, one may assume that it is common knowledge among doctors that safety is their basic professional principle (Annas, 1998). Within the relationship between patient and doctor, the patient may feel placed in the weaker position though, because the doctor has usually the status of expert and the right to trespass barriers of very personal integrity. However, the doctor is not allowed to carry out a medical intervention without the patient's informed consent. This is to ensure that the patient has been informed about benefits as well as risks. Other suitable topics for discussion prior to giving informed consent might be for example how the doctor has organised his clinical practice or his financial relationship with medical device manufacturers (Kessler et al, 1987).

The current regulatory frameworks for health care delivery still focus mainly on the relationship between patient and doctor. Proportional to the ongoing stream of entirely new medical devices and services, the variety of medical professions continues to grow. So does the risk of damage to the patient as medical professions more often behave like competitors than team players. Regulation concerning their competence and behaviour is therefore warranted.

State legislation and self-regulation by participating or interested parties can make a substantial contribution to risk control. However, this is not always the case, as was illustrated in 1993 by an FDA commissioner who declared in his Hastings Lecture that 50 percent of all medical device recalls (including voluntary ones) were due to design shortcomings and 30 percent to violations of the so-called Good Manufacturing Practice guidelines (Fielder, 1993). It turned out that the tone of advertisements and promotion activities had virtually always been far too optimistic and the manufacturer information about performance safety had been incomplete and sometimes even misleading (Cromheecke et al, 2000).

Self-regulation is usually only effective when it concerns a product that is relatively simple to describe and to provide, with widely accepted standards regarding its safety performance (e.g., aviation), or when it concerns a product that is not allowed onto the market unless it has successfully passed various process stages by responding adequately to laws of certainty and predictability and technically engineered solutions (Freed et al, 2001). Medical device manufacturing, medical device application, and patient safety are not such simple issues. Past experience suggests that state regulation is necessary for patient safety, in addition to the industry's own efforts.

SAFETY CONTROL OF MEDICAL DEVICES BY MEANS OF LEGISLATION

State legislation should serve two objectives. First, it should ensure that manufacturers and doctors apply safe medical devices. This is the proactive objective of legislation. Second, it should encourage safe behaviour and damage control by means of sanctions and a duty to financially compensate the damages. This is the reactive or repressive objective of legislation.

The European Union (EU) defines a medical device as follows (Horton, 1995): *"Any instrument, apparatus, appliance material, or other article, whether used alone or in combination, including the software necessary for its proper application, intended by the manufacturer to be used for human beings for the purpose of:*

- *diagnosis, prevention, monitoring, treatment, and alleviation of disease;*
- *compensation for an injury or handicap;*
- *investigation, replacement, or modification of the anatomy or of a physiological process;*
- *control of conception,*

and which does not achieve its principal intended action in or on the human body by pharmacological, immunological, or metabolic means, but which may be assisted in its function by such means".

So, EU legislators regard many products as a medical device, varying from portable home labs (e.g., the Coaguchek™ for self-assessment of anticoagulation) and ECG equipment to pacemakers, mechanical heart valves, artificial ocular lenses, intra-uterine devices, condoms, and intracutaneous insulin pumps.

EU regulations for medical devices are meant to serve the following purposes:

1. improvement of safety and efficacy;
2. stimulation of development and free trade in the EU, and stimulation of export.

For the purpose of improving safety and efficacy, there are three Specific Product Safety Directives: for active implantable medical devices, for other medical devices, and for in-vitro diagnostic medical devices. There is also a General Product Safety Directive, intended for *"all medical devices that are supplied in the course of a commercial activity for use by consumers"* to which none of the three Specific Product Safety Directives can be applied.

Medical devices are further classified according to the level of potential patient risk, varying from low (class I) to high (class III). The requirements for evaluation and certification are coupled to this classification. For class-III devices, for example, extensive laboratory testing, user experience, and patient follow-up are required.

A prerequisite for free trade that still needs to be worked on, is the harmonisation of national legislation on safety, certification, and user circumstances, because the introduction of free trade would mean that all EU-certified devices are allowed onto the market of any member state, irrespective of national regulatory requirements. Member states should also realise that once free trade has been introduced, national legislation on manufacturer product liability will be of secondary importance, in favour of the usually much less stringent EU legislation.

Legislation and proactive risk control

The Product Safety Directives are all fairly technical and succinct. Their objective is to provide member states with the guidance needed to adjust or to generate legislation. The Directives refer explicitly to the safety requirements and performance standards drawn up by the European Committee for Standardization (CEN) and the European Committee for Electrotechnical Standardization (CENELEC).

For the development and adjustment of these requirements and standards, these bodies depend on so-called Technical Committees, mainly consisting of civil servants and of so-called regulatory affairs officers, representing the medical device industry. The medical profession and consumer organisations are also entitled to representation in these committees, but in practice they hardly participate due to insufficient expertise and funding.

Semi-public and private certification bodies may apply to the CEN and CENELEC for the authority to act as a so-called Notified Body with respect to checking data acquisition, file completeness, and compliance with Technical-Committee standards. For class-III devices, the Notified Body may impose on manufacturers the duty to ensure clinical patient follow-up and periodical updates on user experience with a specific device.

In the US, standardisation and certification procedures are generated and enforced by the US Food and Drug Administration (FDA), with the assistance of special FDA panels in which medical experts and consumer organisations are well represented (Fielder, 1993). Successful completion of all five stages occurs under the supervision of a special FDA-certified agency.

Although the US and EU standards for laboratory and animal testing are very similar, investigational device exemption and pre-market approval are far more cumbersome to obtain in the US than in the EU. While the means to develop and to market a medical device safely are abundantly present in the US, the thorough FDA certification procedure hampers its rapid availability to the general public.

Because it is far easier for US manufacturers to obtain a CE (European Committee) mark than FDA approval, they often decide to do the clinical testing in (Eastern) Europe, so that they can start recovering their investments. Other favourable aspects in the eyes of US manufacturers are that in the EU the legal protection of patients included in a clinical trial and the supervision by medical ethics committees are often less restrictive than in the US.

In the EU and US the manufacturing process itself is subject to the so-called Good Manufacturing Practice (GMP) guidelines, which may be considered a form of manufacturer self-regulation . Although one may assume that the content of the GMP guidelines is common knowledge among medical device manufacturers and health care authorities, the required quality-control intensity and extent are often topics of dispute. GMP assessment is part of the certification procedure by the Notified Body. Notified Bodies have the right to demand that the "instruction to users" be translated into a language of their choice and understandable to users of a particular country.

Competition between the roughly 50 Notified Bodies is generally encouraged by the EU. Where certification is denied or where specific conditions must be met prior to certification, the manufacturer has the right to request that another Notified Body be assigned to his case.

Legislation and litigation as tools for reactive or repressive risk control

Once a Notified Body has verified that the content of a certification file meets all safety requirements and performance standards, it will issue a CE mark. This CE mark will carry the special identity code of the Notified Body involved.

The duties of EU manufacturers with respect to adverse-event notification and evaluation have been regulated in the Product Safety Directives, but for the operation of a vigilance system to enhance patient safety, there are only guidelines and no law. In practice, this means that EU manufacturers are obliged to notify any adverse event to the competent authority of their country (e.g., the medical device agency of the National Health Inspectorate). The
competent authority is subsequently responsible for recording, analysis, and notification to other member states.

In the Directives, two categories of adverse events are distinguished:

Category 1. *"Incidents that directly or indirectly might lead to or might have led to death or serious deterioration of the state of health of patients, users, or other persons due to any malfunction, failure, deterioration, inadequate labelling, or instructions in the characteristics or performance of a device." (The meaning of the term "serious deterioration" is to be determined case by case, if possible with the assistance of a medical professional.)*

Category 2. *"Notification is obligatory when a technical or medical reason in relation to the characteristics or performance of a device leads to systematic recall of devices of the same type by the manufacturer."*

State legislation comprises general rulings with respect to breach of contract and tort. Device failure under US civil law and the Safe Medical Device Act (Fielder, 1993) generates strict liability of the manufacturer. A contract between patient and manufacturer is assumed, and the latter is obliged to pay for the damages, including pain and suffering. Where the plaintiff can prove gross negligence or bad faith on the part of the manufacturer, a judge or jury may award punitive damage. This is an extra penalty for the negligent party, intended to punish poor behaviour within the framework of a civil action. Medical device failure is rarely a single event. In the case of structural failures, a series of patients is often affected. They can join together in a so-called class of victims, which institutes at a single court a so-called class or joint action. Apart from the damage to the image of the company, these legal actions have a significant impact and induce an apparent awareness for liability prevention. In other words, civil litigation in the US shifts the burden of proof primarily to the manufacturer, away from doctor, hospital, and patient. Only malpractice or gross negligence have to be proven by the plaintiff. The Safe Medical Device Act also obliges manufacturers to maintain a proper registry of all device users and to carry out thorough adverse-event monitoring. Civil action therefore provides both proactive and reactive consumer protection. In contrast, the EU tort system usually denies patients financial compensation, even in case of failure of an experimental device. The patient has to prove substandard practice, i.e. deviations from the state of the art, or from the state of the industry, or both. Certainly, the rules of breach of contract do not apply, although the patient usually is allowed a claim against doctor, or hospital, or both.

So, in addition to the conventional legislation on product liability and consumer safety, specific medical device legislation is in force in both the EU and the US, but the systems differ. Patient protection in the EU depends mainly on proactive EU legislation and the states.

SAFETY MANAGEMENT BY MEANS OF A PROFESSIONAL SAFETY BOARD

Legislation on safety control in health care usually specifies duties and objectives and ensures facilities for implementation, but not much attention is paid to the management of safety in practice. It is true that the US Medical Device Act prescribes the maintenance of user registries and adverse-event notification to authorities and patients, and EU manufacturers are also obliged to officially report adverse events, but legislation on crisis handling, on solving conflicts of interest, on learning from incidents, and on incident prevention by compliance with the rules is hard to find (Freed et al, 2001).

One has to admit that it is not easy to draw up guidelines for safety management, nor is it easy to put safety management into practice, because when an adverse event does occur, one often has to make fast decisions, on the basis of insufficient data, knowledge, and experience. Within complex organisations such as hospitals, safe behaviour of individuals does not automatically prevent human error, or organisational and intrinsic device failure. Safety has to be well considered and defined in terms of safety standards and safety performance for the various treatments. This is usually left to the professional bodies. Safety management requires specific knowledge, which is usually only available among those professionals. This level of regulation of safety management by professional bodies needs to be embedded closely in existing state regulation as well as in the actual behaviour and practice of the professionals. However, this form of self-regulation often lacks consistency and transparency for the general public.

Given the importance and the complexity of safety management and given the fact that legislation falls short, the involvement of a professional safety board, consisting of heavy-weight professionals with a mission, with authority, with knowledge, and with skills, seems a useful alternative.

In The Netherlands, we already have experience with a professional safety board for the endovascular treatment of the abdominal aortic aneurysm (de Mol, 2000). The aneurysm, in this case a dilated part of the abdominal aorta, often needs to be removed because of the risk of rupture, especially when its diameter continues to increase. Removal used to involve major abdominal surgery, with a mortality rate of approximately 5 percent. This mortality rate applies to a selected group of patients. Many patients with such an aneurysm do not qualify for this treatment, because they are elderly and often not fit enough for major surgery.

In 1993, a new - endovascular - treatment of this aneurysm became possible, which is much better tolerated than surgery, and which is carried out jointly by an intervention radiologist and a vascular surgeon. During the intervention, a tiny catheter with a folded vascular graft is introduced into the

abdominal aorta via the artery in the groin. The vascular graft is subsequently unfolded at the site of the aneurysm, thereby usually solving the problem.
It seems a simple intervention, but apart from the risk of incorrect graft delivery, several problems may occur:

1. Insufficient fixation of the graft above and below the aneurysm.
2. Bleeding between the graft wall and the aneurysm sac, with the risk of vessel wall rupture and graft dislocation.
3. Damage to the graft wall, which is very thin, causing it to be vulnerable. The graft has to be folded inside the catheter and captured in a framework of stents to prevent collapse and to enhance fixation.

Given the fact that the mortality rate of successful endovascular treatment was less than 2 percent in a patient population that also included the elderly and given the fact that abdominal major surgery could be avoided, surgeons and radiologist were inclined to use this technique also on compassionate grounds. However, long-term follow-up revealed that approximately 35 percent of the patients show device failure, requiring surgical intervention. Another set-back was that all devices used for endovascular treatment and marketed in The Netherlands were subject to at least one recall. This was due to causes such as fatigue fracture of the fixation hooks, suture breaks, rupture of the graft texture. There was also lack of follow-up in spite of FDA ruling, and inadequate user instructions and flaws in the application protocols were observed.

Professional Safety Management in Practice: The Safety Board at Work

From the moment the first incidents were reported, the efficacy of the technology was questioned and the need for risk control was considered obvious. The Dutch Society for Vascular Surgery and The Dutch Association of Intervention Radiologists immediately took the initiative to draw up guidelines for this particular procedure in order to improve quality and safety. These guidelines related to requirements for:

- the level of user experience
- the quality of imaging equipment,
- the need for central peer review of protocols for diagnosis and treatment,
- the obligation to maintain a user registry, and
- the need for a professional safety board to analyse adverse-event reports and to advise peers and other interested parties on safety measures and risk communication.

The safety board that was set up, consisted of five people: two vascular surgeons, two intervention radiologists, and one safety expert, who also acted as chairman. From vascular surgeons and intervention radiologists all over the country, the safety board received reports of major incidents, which could lead to liability threats and conflicts with manufacturer and health authorities.

From the other side of the spectrum, the safety board was approached by the Dutch health authorities for consultation, for example on the interpretation of tests conducted by the manufacturer, and for comments on proposed diagnostic and therapeutic interventions and risk communication strategies. In two cases, the safety board advised negatively when asked by the health authorities for its opinion on the continuation of use of a device that was already marketed, and a redesigned device, both produced by the same manufacturer. The safety board did not have the authority to withdraw devices from the market, and the Dutch health authorities did not want to do it, due to the risk of litigation. However, these two negative pieces of advice, which were relayed to doctors, led to device withdrawal from the Dutch market, an example which was followed in many other EU countries, and finally world-wide. Meanwhile, the safety board continues to be briefed on new incidents and developments with respect to close-monitoring programmes for patients, and to be asked for comments.

In The Netherlands, this professional safety board for the endovascular treatment of the abdominal aortic aneurysm has acted as a model. The introduction of a professional safety board for many other health care devices and procedures is now generally advocated by safety experts and health authorities for the following reasons (Wood et al, 1998):

1. A safety board can proactively design safety measures on the basis of a safety analysis and guidelines at a moment when all parties involved still have in interest in prevention and control of safety risks. After an incident, conflicts of interest may occur, which may interfere with the execution of adequate safety measures.

2. The safety board can encourage compliance with self-imposed measures such as peer review of protocols, adverse-event reporting, and registration enrolment. The board can advise in cases of risk communication and recalls.

3. The safety board consists of experts with experience in retrieval analysis and technical research. The board call upon specific knowledge and skills of independent experts to decide upon the parties to be held accountable and required to execute the risk control procedures.

4. Prior to large-scale device use, safety is often assumed and estimated on the basis of assumptions and surrogate endpoints. The safety board can demand verification and extra tests where legislative regulatory standards are considered inadequate.

After adverse-event notification, the safety board will act as follows.

Step 1. Incident analysis. The safety board focuses on matters such as aetiology, prevalence, recurrence, and possibilities for monitoring and improved follow-up, and it reviews the information provided by manufacturer and health authorities.

Step 2. Risk management. The safety board focuses on matters such as early detection, intervals for check-ups, independent expert assessment of the X-ray checks, and facts supporting a communication strategy to patients and media. Monitoring and early detection make sense only where a life-threatening situation can be handled adequately. The safety board encourages manufacturer and doctors to agree on a common risk management strategy.

Step 3. Accountability of parties and potential legal consequences. All plans and intended actions have to be carried out by authorised persons. The safety board may separate intertwined interests by clarifying the various duties. Manufacturers often have professional communication and regulatory staff on hand in times of crises in order to defend their interests. Making the first move or setting the scene is very important in case of expected public outrage. For matters relating to patient safety, including the transfer of manufacturer information, patients should, however, be approached by their doctor. The assistance of patient interest groups and/or patient safety movements may be valuable (Brewin, 1994).

In summary, because legislation falls short in terms of dealing with safety management, the introduction of professional safety boards within the framework of self-regulation seems attractive. However, health care professionals are not always interested in this form of self-regulation , and it takes time to draw-up guidelines and agreements and to gain authority.

Considering the high stakes, the complexity of safety management, and the fact that the basic relationship between doctor, patient, and manufacturer is regulated by law, it might be preferable to incorporate safety management into the legislative system for health care rather than leaving it to professional self-regulation .

EVALUATION OF THE REGULATORY FRAMEWORK FROM A PATIENT PERSPECTIVE

Communication between professionals and between professionals and patients determines retrospectively whether certain behaviour will be considered unsafe (Vincent et al, 1994). Evaluation of the effectiveness of the regulatory framework from the perspective of patient safety in the case of medical device failure is complicated by the following factors:

1. A definition or classification of medical device failure is missing.
2. The number of adverse events appears to be irrelevant. In daily practice few medical device failures lead to patient damage. They are rarely encountered within the same hospital and therefore often experiences as bad luck. There is no obligation for doctors to report device failure, which often remains unrecognised due to lack of experience of how to assess and to confirm a typical device failure.
3. It can be difficult to determine whether device failure is disease-related, patient-related, or doctor-related, rather than device-related, and whether it has led to patient damage.

If a complete quantitative medical device failure registry were to be available, much could be learned from it in terms of avoiding medical device failure, because it would allow modelling for early warnings, quick ratings, predictions, and cost estimations. Based on the outcome, patients could be informed adequately about risks and benefits before agreeing to a certain treatment. As mentioned earlier, informed decision-making and informed participation are in fact basic patient rights. In heart surgery, the importance of observing these patient rights is always recognised. When a patient has to undergo heart valve replacement, for example, he may have the choice between a biological valve and a mechanical one. The first is known to have a limited durability (giving rise to the risk of re-operation) but no anticoagulation therapy is required, whereas the second is known to have a prolonged life, but anticoagulation therapy is essential (giving rise to the risk of bleeding or clot formation in the case of badly controlled anticoagulation).

The US proposed bill of patient rights goes a few steps further: in addition to the right of informed consent, patients have the right to know the surgeon's identity, success rate, and clinical arrangements in terms of finances and practice organisation (Kessler et al, 1987). The objectives of all of these rights are that the relationship between patients and doctors is characterised by mutual trust and that patients feel confident that they have made the best possible decision, and remain so even in the case of negative treatment outcome.

To convince the public that the safety of medical devices in general, and of certain devices in particular, is sufficiently guaranteed, the following conditions must be met:

1. The risk of failure of medical devices in general and for each device in particular should be considered acceptable. When good performance and absence of adverse events can be predicted with sufficient certainty, the outcome can be discussed with the patient in order to allow him to balance risks and benefits.

2. The type of failure and the failure scenarios should be considered acceptable. Failure due to lack of knowledge in the design stage and in the choice and development of technology may be considered acceptable. Failure due to wrong decisions with respect to design and testing of a

device, manufacturing shortcomings or shortcomings in the certification procedure are generally not considered acceptable.

3. The type of patient damage and the size of material and societal costs should be considered acceptable. Death and disability are considered less acceptable in young patients than in elderly patients, for example.

4. A regulatory framework in order to avoid and to control failure and possibilities of litigation should be in place. Data should be available on the comparison of risks and hazards of technologies and services of similar nature.

The level of public outrage following device failure and failed risk control is determined by a number of aspects: lethality, scale and chances of survival, and risk perception (i.e., voluntary risk versus imposed risk). A lack of fairness on the part of a perpetrator and a lack of social equality for a victim increase public outrage (Sandman 1991).

Public outrage can have substantial impact on politicians and other public figures responsible for legislation and enforcement. This was illustrated in the 1990s, when the Björk-Shiley scandal due to fractured mechanical heart valves led to major changes in the US regulatory medical device system (de Mol et al, 1995).

THE AETIOLOGY OF A DEVICE SAFETY PROBLEM IN THE YEAR 2000

Legislation does not always lead to adequate safety management and proper patient protection. In fact, the certification process may be dominated by economic interests while neglecting patient safety, as the following example will illustrate.

Around 1995, the world's biggest manufacturer of mechanical heart valves, St. Jude Medical, decided to apply elemental silver on the texture of the valve's sewing ring (Tozzi et al, 2001). The company did so because silver has the property of killing microbes, and it assumed that the application would reduce the rate of prosthetic valve infections, an often-fatal complication of heart valve replacement. This rate is on average roughly 2 percent in the first postoperative year and roughly 0.5 percent in the years thereafter.

In its request for an abbreviated approval procedure in order to get access to the market, the manufacturer presented the silver application to the certifying bodies as a minor adjustment, not affecting any of the valve's mechanical properties. The FDA, convinced of the benefits, granted the request, asking only for laboratory testing with a focus on the efficacy and safety of the silver application. The required testing subsequently confirmed that the silver application killed most microbes and that it had no toxic effect on tissue and blood. Based on these findings, the FDA ruled that the so-called Investigational Device Exemption (IDE) and Pre-Market Approval (PMA) were not necessary in this particular case, and granted the manufacturer permission for marketing and clinical use within the US. A CE-certificate, for marketing and clinical use in the EU, was also readily provided.

For commercial purposes, the manufacturer then took the initiative to arrange for a prospective randomised multi-centre clinical trial, comparing patients receiving a valve with the silver application with patients receiving a conventional mechanical valve. Due to the low incidence of prosthetic-valve infection, large patient numbers, involving many clinics, were required to prove the anticipated improvement. Meanwhile, surgeons also started to opt for the new valve for non-study patients who suffered from native-valve infection, because they (wrongfully) assumed that the silver application would also be beneficial in case of existing infections, as shown in in-vitro testing.

When the clinical trial had been underway for two years, routine echocardiography in study patients revealed that in the group with the silver application, valve leakage was eight times higher than in the other group. The silver apparently not only killed microbes but also heart tissue, thereby preventing prosthetic-valve ingrowth, thus causing valve leakage.

It was sheer luck that the increased rate of valve leakage, which usually only gradually leads to heart failure, was diagnosed in such a relatively early phase, especially because large prospective clinical trials involving a new prosthetic heart valve had not been carried out earlier. Moreover, the new valve had been implanted mostly in patients not enrolled in the trial. Naturally, the process of valve leakage was subsequently watched closely by means of echocardiography, the valve was replaced if necessary, despite the increased complication rate of redoing the valve replacement, and the valve was withdrawn from the market.

Retrospectively it was noted that other medical disciplines had already banned the use of silver much earlier because of its toxicity under physiological conditions. However, the panels involved in St. Jude Medical's request for certification had considered this expertise irrelevant in their battle with the manufacturer's technical representatives. The FDA had considered the silver application on the sewing ring a minor improvement, which, based on the certification requirement system, did not require animal testing or small clinical patient trials before large-scale marketing.

Furthermore, although randomised clinical trials are effective to prove or to reject a claim for improved prosthetic-valve performance, neither the FDA nor the EU certification regulations demanded such validation, a policy which has remained unchanged for similar cases to the present time (Antman et al, 2001).

From the patient's viewpoint, medical devices are considered safe if the following technical and regulatory requirements have been met:

1) good design and manufacturing specifications, correct indications for use, proper use by medical professionals, effective maintenance and support, and adequate patient follow-up, allowing timely correction of evolving failure;

2) appropriate testing standards, user certification by independent authorities, known accountability, risk anticipation, risk control in case of failure, and right of litigation.

As mentioned earlier, state legislation rather than self-regulation should provide the basis for safety management and patient safety.

WHAT MAKES FOR A BETTER REGULATION OF SAFETY IN MEDICAL DEVICES?

Three levels of safety and risk control for health care are served by three types of regulatory framework: 1) state legislation, both general and specific for health care ; 2) medical professional organisations and user groups; 3) patient-doctor relationship. The prerequisites for each of these levels differ and they therefore need to be addressed separately.

Prerequisites for medical legislation (level 1)

1. Objectives should be clear and societal support should be broad. The EU system of Directives is diffuse, because it serves free trade as well as patient protection. Also, the system is far clearer with respect to serving economic interests than with respect to serving patient safety.

2. Codification should be elaborated and effective. EU national laws apply to both national legislation and rules for primary processes, but they fall short in terms of dealing with safety management. So far, only a few sets of specific guidelines have been drawn up, such as the Good Manufacturing Practice guidelines, but, as mentioned earlier, this is a form of manufacturer self-regulation.

3. Safety standards should have objectives that are clear and easy to apply. EU legislation refers mostly to the safety standards set by the so-called Technical Committees, without clarifying the objectives. The safety standards do not focus on durability, success rate, or number and type of adverse events.

4. Requirements for certification and re-certification should be defined. Certification has been regulated with respect to the properties of a medical device before it is allowed onto the market for clinical use, but there are no requirements for feedback based on post-marketing surveillance and retrieval analysis. Only product recall and withdrawal from the market in the case of a serious adverse advent have been regulated.

5. An active inspection body should be available. Contrary to what is common practice in other industries, on-site inspection visits to check compliance of medical device manufacturers and medical professionals are rare. The requirement is also seldom included in frameworks for self-regulation .

6. Legislation should be enforced. In the case of violations, legislation provides ample opportunities to correct and to sanction, based on common law, employment law, civil law, and/or criminal law. One should realise, however, that sector-specific situations require sector-specific actions and measures in order to correct and sanction sensibly and effectively.

Prerequisites for medical professional organisations and user groups (level 2)

1. Definition of safety improvement cycle. It appears to be difficult for professional groups to distance themselves from the interests of official bodies that supports them. Hospitals acts as employer, health care insurers as judges whether the use of a device can be reimbursed, and industry often acts as a sponsor. Therefore, a priori standards of performance and type of failure for a certain generation of devices should set in advance.
2. Communication. Absolutely safe devices are unaffordable and not available. Device failure will occur despite all efforts due to lack of knowledge. These failures are often magnified by competitors or interest groups. The first response of the manufacturer and doctors involved is often a cover up and an "optimistic and reassuring version" of truth at the expense of effective safety measures. Adequate communication with the public and the potential victim is mandatory, as a patient who knows is able to protect him self best (Cromheecke et al, 1999).
3. Willingness and ability to learn. Incidents often result in a recall. Similar products move into the vacuum, although their safety level is still not established. A recall should not only warn the sector, but also force it to consider safer design or manufacturing practices.

Prerequisites for regulating the patient-doctor relationship (level 3)

1. Mutual trust. One must realize that the quality of a patient doctor relationship can not be determined in advance by means of laws or institutional protocols and guidelines. However, it is still a part of the craftsmanship to serve patients well and fair. When lack of trust and inability to communicate result in poor performance related to a treatment by means of a device, the primary principle of patient protection require the existence of very specific measures. This may result in prohibiting a doctor from using that particular device or treating a certain type of patients.
2. Informed consent. Medical treatments always carry a risk. The assessment whether such a risk should be considered acceptable is largely determined by the information provided prior to treatment.
3. Recognition of patient rights to safe treatment and compensation in case of damage. Today, many manufacturers and doctors experience device failure as a serious threat to their professional credibility. They therefore tend to deny that the device failed or, alternatively, that the failure was due to any short-coming in the device. This attitude prevents learning from failures and denies justice to patients.

CONCLUSIONS AND RECOMMENDATIONS

Safety management in health care is a process that should have the objective to ensure medical device safety and patient protection as a mission. It should be embedded in a regulatory framework based on legislation. It is, however, not yet covered by current regulation, which mainly consists of detailed rules to obey.

The main objective of the medical device industry is the profitable marketing of a technically safe product. The industry's proactive regulations lack specification of safety standards. Considering the concept of balanced interests, medical device safety and patient protection are marketing points rather than fundamental goals. The industry's reactive regulations provide guidelines on how to deal with accidents and on how to deal with compensation claims in case of damage, rather than guidelines on how to learn from accidents in terms of prevention. Risk communication in the medical device industry mainly serves internal-control purposes and market share protection. Litigation is considered a nuisance and a threat to the whole medical device industry.

Based on the shortcomings with respect to safety management revealed by recent medical device failures, involving heart valves, pacemakers, and breast implants, we would recommend that the following be undertaken:

1. Development of guidelines on how to deal with the trade-offs between patient safety and economic interests, and of management guidelines for conflicts of interest between medical device manufacturer, doctor, and patient.

2. Inclusion of the implantation and follow-up phases of the product life cycle in the risk control process. This would require safety management systems, which are in general not yet available in the medical device industry and health care.

3. Legislation and regulatory frameworks that provide tools for effective improvement of patient safety or effective enforcement of safe conduct by medical device industry and medical profession.

4. Information with respect to risks, specifications about performance, and transparency of how a medical device is selected, available for controllers in the system as well as for patients.

5. Effective risk control and safety management by meeting the following criteria:
   a) all safety measures should be applied simultaneously;
   b) risks are verified by experience;
   c) the originally certified design is used;
   d) adequate commercial management is matched with a culture that respects safety;
   e) conflicts of interest and of authority are dealt with;
   f) risks of failure are accepted in a publicly credible way.

6. Eradication of justifications for violation of standards for safety, ethics, and professionalism, such as:
   a)"Earn as you learn". New technology, often a me-too variation, implies uncertainty and risks. Under these circumstances, decision-makers consider a commercial trade-off with safety costs acceptable. The ultimate test of a medical device is patient use.
   b)"Higher / public interests are being served". This excuse may be used to preserve the credibility of politicians responsible for the functioning of "competent authorities" but also to guard against loss of market share or the consequences of a bad corporate image of a country's high-tech branch.
   c) Avoidance of employee lay-offs and recovery of investments by small-pocket shareholders.
   d) Potential loss of research & development money, especially when a substantial part was public money.
   e) Salvaging of reputation.

Although regulatory frameworks encourage individual implementation of safety measures, connected regulation at the three levels result in transparency, clear duties and standards, and will promote a level of safety that will pay off for the whole industry. One of the benefits of litigation is that it allows for impartial assessment of the regulatory framework, thereby revealing whether regulation might require repair or redesign.

Another benefit is that especially civil litigation may capitalise the costs of the lack of safety and accountability, and clarify just what is "state-of-the-art" and "state-of-the-industry". So, although litigation is a burden, also promotes progress.

REFERENCES

Annas, G. (1998). Bill of patient's rights. *New England Journal of Medicine* 1998;**338**:695-9.

Antman, K., S. Lagakos, J. Drazen (2001). Designing and funding clinical trials of novel therapies. *New England Journal of Medicine* 2001:**344**:762-3.

Brewin, Th. (1994). *Primum non nocere?* Lancet 1994;344:1487-8.

Cromheecke, M.E., A.P. Yazdabbakhsh, B.A. de Mol (1999). Liability for failed mechnical heart valves: an accident and risk perception analysis. *International Journal of Risk & Safety in Medicine* 1999;**12**:81-7.

Cromheecke, M.E., F. Koornneef, G.L. van Gaalen, B.A. de Mol (2000). Controlling risks of mechanical heart valve failure using product life cycle-based safety management. In: Vincent Ch, B.A. de Mol (Eds). *Safety in medicine*. Elsevier Pergamon, Oxford, 2000, pp 1-21.

de Mol, B.A. (2000). The safety of endovascular treatment of aortic aneurysms. In: Branchereau, A., M. Jacobs (Eds). *Surgical and endovascular treatment of aortic aneurysms*. Futura Publishing Company, Inc. Armonk, NY, 2000, pp 307-15.

de Mol, B.A., F. Koornneef, G.L. van Gaalen (1995). What can be done to improve the safety of heart valves? *International Journal of Risk & Safety in Medicine* 1995;**6**:157-68.

Fielder, J.H. (1993). Getting the bad news about your artificial heart valve. *Hastings Center Rep* 1993;**23**:22-8.

Freed, C.R., P.E Greene, R.E. Breeze, W.-Y. Tsai, W. Du Mouchel e.a. (2001) Transplantation of embryonic dopamine neurons for severe Parkinson's disease. *New England Journal of Medicine* 2001,**344**:710-9.

Gutheil T.G., H. Bursztajn H., A. Brodsky (1984). Malpractice prevention through the sharing of uncertainty. *New England Journal of Medicine* 1984;**311**:49-52.

Horton, L. (1995) Medical device regulation in the European Union. *Food & Drug L J* 1995;**461**:463-4.

Kessler, D.A., S.M. Pape, D.N. Sundwall (1987). The federal regulation of medical devices. *New England Journal of Medicine* 1987:**317**:357-66.

Kohn, L.T., J.M. Corrigan, M.S. Donaldson (eds) (1994). *To err is human. Building a safer health system.* Committee on quality of health care in America. Institute of Medicine. National Academy Press, Washington, DC, 2000.

Leape L.L., Brennan T.A., Laird N. et al (1991). The nature of adverse events in hospitalized patients. Results of the Harvard Malpractice study II. *New England Journal of Medicine* 1991;**324**:377-84.

Sandman, P.M. (1991). Emerging communication responsibilities of epidemiologists. *Journal of Clinical Epidemiology* 1991;**44**(suppl.I):41S-45S.

Tozzi, P., A. Al-Darweesh, P. Vogt, F. Stumpe (2001). Silver-coated prosthetic heart valve: a double bladed weapon. *European Journal of Cardio-thoracic Surgery* 2001;**19**:729-31

Vincent, Ch, A. Young, A. Philips (1994). *Why do people sue doctors? A study of patients and relatives taking legal action.* Lancet 1994;343:1609-13.

Wood, A.J., C.M. Stein, R. Woosley (1998). Making medicines safer. The need for an independent drug safety board. *New England Journal of Medicine* 1998;**339**:1851-4.

Baram's paper discusses another regulatory frontier, the control of biotechnology, in particular, new gene therapies and genetically modified crops. He argues that there is, as yet, no coherent regulatory framework but an emerging ad hoc patchwork of controls driven by various public pressures. There are many similarities here with the analysis provided by de Mol and van Gaalan of developments in medical device technology.

Baram notes that the presumption has been that biotechnology was an area where information about failures would be rapidly communicated and that the processes of organisation learning would ensure safety improvements. This was an area, in short, where self-regulation might have been expected to work. He argues, however, that this has not been the case and that commercial pressures and conflicts of interest have prevented the dissemination of critical information. The picture he paints is akin to the wild west frontier where individuals pursue their own interest with little or no control by government.

Baram raises the interesting question of whether the experience of managing other more mature technologies might be of use in achieving more effective social control of biotechnology.

13

BIOTECHNOLOGY AND SOCIAL CONTROL

Michael Baram

INTRODUCTION

Biotechnology is rapidly progressing from laboratory research to the development, testing, marketing and use of a multitude of new products and methods in health care and agriculture. Promoted by substantial investment and incentives for societal benefits and financial gain, this progression has been relatively unimpeded by stringent regulation or other restrictive social controls. However, applications of biotechnology are stimulating deep moral concerns and producing growing evidence of potential harms to health, safety and environment. As a result, public pressures have intensified for restrictive policies and regulations, and federal officials in the United States are now deliberating the steps to take. The paper surveys the main features and causes of this rapid progression, and the emerging patchwork of social controls, focusing on the safety of new gene therapies in health care and genetically modified crops in agriculture. It finds that use of an organizational learning model to protect patient safety in clinical testing of new gene therapies has failed for several reasons, and that performance requirements for the approval of new crops have been too permissive and lack public credibility. It concludes with the perception that proposed regulatory reforms will in the aggregate, ultimately require health care and agricultural organizations to develop and implement safety management systems for more effective self-regulation. This poses a major challenge for these organizations, but raises the possibility that experience gained in managing the safety of other technologies may be of considerable value in achieving more effective social control of biotechnology.

1. THE ADVANCE OF BIOTECHNOLOGY

New technology is a powerful force for change in human affairs and the natural environment. It challenges established interests and creates opportunities. As its applications are managed through regulation and other social controls, it is adapted and absorbed over time by society. So it has been with chemical and nuclear technologies, and new modes of transport and communications, throughout the 20th century.

As we enter the 21st century, biotechnology is advancing upon us at an incredible pace, generating new knowledge about the genetic makeup of all life forms, and providing techniques for modifying them and creating new species. It is also enabling the development of new products and methods for substantially improving human health and agricultural productivity. Thus, biotechnology is rapidly progressing from laboratory research to the development and testing of prototypes, and thence into commercial application in health care and food production systems (Caulfield and Williams-Jones 1999).

This rapid progression is occurring despite intense concerns and disputes about the social and risk implications posed by biotechnology. As with other technologies, many issues relate to managing risks: for example, whether the risks of testing a new genetic therapy on human subjects should be more stringently controlled, or whether more extensive testing of a new genetically modified crop should be done prior to agricultural use in the open environment or prior to sale for human consumption.

But unlike other technologies, biotechnology raises more fundamental issues. One is its potential for use in ways that disturb deeply held moral convictions, such as combining genetic material from unrelated species to create new life forms, for cloning humans or, facilitating eugenic policies. Another is that we really don't know if the unique genetic configurations created for medical and agricultural applications bear any intrinsic hazards which, under variable conditions of use and exposure, would be transformed into risks to health, safety or the environment. Thus, scientific uncertainly about hazard makes discussion or concern over risk quite speculative, so much so that risk regulators claim they lack a sufficient factual basis for imposing prescriptive standards on biotechnology (Rifkin 1998).

This contrasts with chemical and nuclear energy technologies, for example. These technologies have not demonstrated such high potential for moral conflict, and pose known hazards such as toxicity and flammability, ionizing radiation and mutagenicity. Thus, societal concerns and disputes were generally confined to incremental issues of how to appropriately manage known risks. However, genetic methods of classifying people, modifying and creating life forms (human, plant, animal) and designing new products and methods for radically transforming health care and agriculture do indeed create moral conflicts and present significant scientific uncertainty as to whether they could threaten health, safety or environment.

As a result, social control of biotechnology through regulation, for example, is confounded by issues of morality and scientific uncertainty, and is therefore more problematic than for other technologies. It is further confounded by the multiple manifestations of biotechnology to be addressed, i.e. its numerous and diverse products and methods which are now being exploited across society. Again, this contrasts with other technologies where social control is readily aimed at the design and operation of a limited number of physical facilities or discrete systems, or at the manufacture and use of a narrow range of products.

For such reasons, a coherent framework of social controls for biotechnology does not exist, and this circumstance, coupled with strong economic and other incentives for exploiting genetic knowledge, enables its rapid progression.

2. TECHNOLOGICAL AND ECONOMIC FEATURES OF BIOTECHNOLOGICAL ADVANCES

Major research programs are in advanced stages of mapping the "human genome", the entire set of DNA found in virtually every cell of the human body, and the genomes of various species of animals, microorganisms and plants. The vast amounts of raw data produced are being analyzed to identify the interconnected sets of DNA which constitute genes (of which somewhere between 30,000 and 120,000 are believed to comprise the human genome), and the genes which constitute chromosomes (of which 23 pairs comprise the human genome). Researchers are also investigating what selected genes and gene fragments express and the functions they serve, in order to determine their influence on the properties or characteristics of humans and other life forms (*Human Genome News* 2000).

Concurrently, other researchers are developing methods of testing humans and other life forms to determine if they carry certain genes known to be predictors of future illness in otherwise asymptomatic persons and their offspring. Ability to develop such a "diary of one's future health" far exceeds conventional medical testing for diagnosing health status over time. Genetic test methods are being reduced in practice to the point that "home test kits" are being sold in England to enable women to determine if they have the BRCAI gene which predisposes carriers to breast cancer. The number of genetic tests for determining an individual's predisposition to various diseases is growing rapidly, with many being capable of identifying latent diseases for which no therapeutic interventions are known, (e.g., Alzheimer's disease). In addition, large-scale testing programs are being designed for populations with well-documented family medical histories, as in Iceland and Sweden. The correlations between test results and family medical problems documented in the historical records that will be found will facilitate determining the genetic basis for various hereditary diseases and the commercial development of drugs and therapies tailored to eliminate or treat each individual's disease (*Icelandic Health Care Database*).

Accompanying these developments is the perfection of means for manipulating genes and altering their influence. Using recombinant DNA (rDNA), "a kind of biological sewing machine that can be used to stitch together the genetic fabric of unrelated organisms," the "splicing, recombining, inserting and stitching" of new living material is being practiced. This is enabling commercial production of genetically-modified biologic materials for regenerating human tissue (and soon for regenerating entire organs which had become diseased or impaired), genetically-modified bacteria strains to combat insect pests and to destroy hazardous wastes, and a growing menu of genetically-modified crops, fish and livestock (Rifkin 1998).

Greatly expediting such biotechnological progress is the emergent field of bioinformatics. Broadly defined, bioinformatics involves the development and application of computational algorithms and advanced data processing to store, structure and analyze vast amounts of genetic information in order to determine gene structure and function, and to design new gene therapies and drugs for each individual's condition, crops and bacterial strains with specific new features, and other unique products for use in health care and agriculture (DeLisi).

Bioinformatics will similarly aid in the development of new national programs now being initiated in the U.S. These include the "environmental genome program" to determine which genes predispose humans to pollution-triggered illnesses. Thus, employers will be able to determine which job applicants or workers are more vulnerable to workplace chemical risks because of their genetic makeup, a development which will influence the employment prospects of currently healthy but vulnerable persons when employers seek to avoid future harms and costs due to worker illness. Another is the "biofeedstocks program" which would "bioprocess" agricultural materials to produce chemicals and thereby displace use of petroleum feed stocks, a development which could replace an environmentally problematic industrial process with potentially cleaner methods (Sharp and Barret 1999; Franz 2000).

Economic resources, incentives and opportunities have also played an important role in promoting the rapid progression of biotechnology into the commercial marketplace. For two decades, the U.S. government, and to a lesser extent, the national authorities of several other industrial nations have provided billions of dollars for basic and applied research in universities, medical research centers, and national laboratories to map the human genome and the genomes of other species, as well as conduct studies on interpreting and putting genomic data to practical use. More recently, as commercial prospects have become apparent, companies and other private sources have invested heavily in applied research, the development and testing of prototype products and methods, securing government approvals, and marketing, distribution and sale of end products and methods. Profit-seeking entrepreneurs have also played a crucial role in linking the government-funded academic and scientific culture of genetics research to the commercial, profit-seeking drug and agrochemical companies and their world of financial and competitive pressures (Caulfield and Williams-Jones 1999).

The profit incentive has been enhanced by several important and controversial decisions by courts in the U.S. on the patentability of various developments in biotechnology. By interpreting U.S. patent law broadly, the courts have held, for example, that patents can be awarded for the genetic "invention" of novel life forms not found in nature (e.g., genetically-engineered crops, organisms, animals, tissue, organs, etc.); for the isolation of genes and cell lines and the processes for altering them; and for algorithms executed by a computer which manage data and aid in the design or implementation of a functional, physical system (a computer-implemented process) (Merges et al, 1997; Grubb, 1999).

Award of such patents confers upon the individual or organizational "inventor" an exclusive right to exploit the "invention" for economic or other purposes (e.g., to use it for profit or license others to use it in return for royalties) and to exclude others from such use. The astounding breadth of such patents offers substantial economic opportunities. But it can also obstruct others from important research, a circumstance which has led many scientists to join others who decry the patenting and commercialization of life and nature, in opposing such trends in U.S. patent policy (Heller, 1998).

The profit incentive has also caused large American and European companies to aggressively capture knowledge from native cultures in undeveloped regions about medicinal plants and crops with special characteristics. The native populations are also being tested to determine the genetic bases for their characteristic diseases and immunities. By applying genetic analysis to such information, many companies have developed capabilities for commercially exploiting the "folk wisdom" or indigenous knowledge of native cultures and learned how to make derivative drugs and other useful products; have secured patents for exclusive use of such products, and now manufacture the products for their profit. Although often condemned as "biopiracy" or "biocolonialism" because the native sources do not fully realize the potential of what they are contributing or receive commensurate benefits in many instances, there are no effective restraints imposed on these activities by international authorities or the countries where such companies are based (Rifkin 1998).

Thus, remarkable technical capabilities coupled with strong economic incentives and competitive pressures promote the rapid progression of biotechnology. Although a coherent framework of social controls for addressing the moral and risk implications of this progression has not been constructed by individual nations or international organizations, disparate and fragmented social controls have been emerging on an ad hoc basis. In this regard it is instructive to assess the evolution of such social controls in the nation which leads the biotechnology enterprise, the United States.

3. The Evolution of Social Controls

a. Background

By any measure, the United States leads other nations in promoting and using advances in biotechnology. In its tradition of providing large scale financial support for new technologies deemed important for military or economic reasons (e.g., nuclear weapons and energy systems, aerospace, computers and communications, etc.), the U.S. government has provided billions of dollars for mapping the human genome and a multitude of other basic and applied research efforts in biotechnology.

The public has supported government funding because early genetic research, done in academic and medical research centers, was viewed as ultimately being of therapeutic value in health care, and as an enterprise not dominated by corporate profit seeking. In addition, there was little concern that such research would ever be used to create new forms of discrimination or eugenics programs in the U.S., such as Europe in the Nazi era had experienced. Nor was it expected that laboratory-based biotechnology research would lead to risks to public health, safety or the environment. Finally, there was public confidence that the academics and medical researchers doing biotechnology would be responsible and self-regulate to prevent inappropriate outcomes, as evidenced by the "Asilomar moratorium" on genetic research, a self-imposed freeze on such research by geneticists for a brief period in the 1970's until the federal government could develop guidelines for the safe conduct of such research.

Given public support, confidence in the research community, growing national antagonism to restrictive federal regulation, and a strong interest in fostering a technological leap forward with potential economic and social benefits, federal officials in the mid-1980's deliberated the need for national policy to govern biotechnological advance and rejected enactment of a new federal law and regulatory program for controlling biotechnology. As a result, a laissez-faire policy was developed by the President's Office of Science and Technology Policy (OSTP 1986; 1992).

OSTP, through a series of official policy statements issued by the President, pronounced that biotechnology should be fostered by federal agencies, that the agencies should use pre-existing regulatory mandates (for medical product and food safety, environmental and workplace protection) in enacting any risk regulations, and that any such regulations should not discriminate against biotechnology advances because of their genetic features. Thus, the genetically engineered drug or food, for example, would only be regulated according to proven risks and not be subject to special restrictions because of the genetically engineered method by which it was produced. No attempt was made by OSTP to delineate any moral or ethical parameters or designate who would address such issues (OSTP 1986; 1992).

Through the 1990's, Congress and the President continued to provide substantial funding for biotechnology, and risk-regulating agencies refrained from enacting restrictive requirements. Certain agencies even began waiving some of their precautionary requirements for the pre-market testing of genetically modified crops.

b. Health care sector

However, as biotechnological advances during that decade began to raise moral and ethical issues which transcended agency mandates for risk regulation, public pressures intensified, demanding a federal response. As a result, Congress and the President have responded on an ad hoc basis to outbreaks of public outrage by, for example, restricting the use of fetuses in genetics research, preventing the use of federal funds for efforts related to the cloning of humans, establishing advisory boards to develop guidance on genetic testing and other matters, and exhorting against patenting the raw data generated by the federally-funded "human genome" research program.

Thus, a patchwork of official constraints based on moral and ethical concerns has started to emerge, with more to come. Congress is considering enactment of laws that would enable persons undergoing genetic testing to control dissemination and use of test results in order to protect their privacy and prevent health insurers and prospective employers from gaining and using such information in ways that discriminate: i.e. to prevent insurers and employers from denying insurance coverage or employment to persons whose genetic tests indicate a predisposition to illness or behavioral problems. Arrayed in opposition to this measure are researchers and companies who want to ensure easy access to such test results in order to advance their professional and economic interests (Baram 1997).

Many other ethical issues regarding genetic testing must also be addressed in the near future: for example, whether a person's test results should be conveyed to spouses, partners or families when such information would help them protect their health, whether all infants should be genetically tested for certain hereditary diseases in order to enable preventive or remedial interventions, and numerous issues regarding the rights of special test populations in the military, in prisons, and in mental institutions (Secretary's Advisory Committee on Genetic Testing, 2000).

Evidence has recently emerged that certain applications of biotechnology endanger health, safety and the environment. In the health care context, the death of a young patient enrolled in a clinical trial of a gene therapy designed to cure serious metabolic disorders occurred in 1999. Investigations established that the death was due to the gene therapy, that the patient had suffered an overwhelming immune reaction to the viral vector used in the trial to "infect" patient cells with new genetic material. Since thousands of patients are enrolled in some 400 other gene therapy trials, many of which use the same type of viral vector for implanting genetic material in patients, the agencies which fund, authorize and oversee these experiments (the National Institute of Health and the Food and Drug Administration) suspended many trials and broadened their investigation in order to determine if other patients were at risk (Testimony of Amy Patterson, 2000; Bush, 2000).

The full scale investigation is still in progress but has already produced findings which have shaken public confidence in medical research on human subjects, the presumed safety of gene therapy test procedures, and the watchdog agencies. It has been determined that some 500 "adverse events" were suffered by patients, including up to 10 deaths; and that the vast majority of these events had not been fully reported to the agencies by the researchers or their institutions (medical research centers), a violation of agency requirements. Further investigation is being done to determine if these events were caused by the gene therapies, or as researchers claim, by other causes, and to also discern why reports had not been made.

Other findings indicate that many researchers and organizations failed to comply with clinical protocols for carefully assessing and selecting patients for the trials and for monitoring them thereafter. Additionally, it was found that many doctors had failed to fully inform prospective patients of uncertainties about the benefits and risks of the therapies or other relevant information (e.g., that other enrolled patients and test animals had suffered adverse events). Thus, doctors had enrolled patients without their "informed consent" to medical experimentation in violation of clinical protocols and medical ethics.

Perhaps the most disturbing finding is that many of the medical researchers and their organizations have substantial financial holdings in the biotech companies which make the genetic materials being tested by the researchers on their patients. Thus, they stand to benefit from speeding the trials to conclusion, having a readily available pool of ill-informed human subjects, publicizing positive results and refraining from disclosure of problems such as adverse results. Their benefit, in the form of return on financial investment, arises when sufficient trials are speedily concluded with good results because agencies will be more likely to approve the therapies and materials for widespread medical use, enabling the companies to do large scale production and sale of the materials, profit therefrom, and attract additional investments. As a result, the holdings of the researchers and their organizations would likely increase in value, and they would also be more likely to be awarded further grants for continuation of gene therapy research (Kowalczyk, 2000; Green, 2000).

Evidence that reports of adverse events have not been made, and that medical researchers and organizations have such financial interest in the trials, combine to create the inference that patient safety is secondary in gene therapy to financial gain. It also appears to demonstrate that social control, in the form of regulatory use of an organizational learning model for ensuring patient safety (Vincent and de Mol, 2000) in gene therapy trials has failed. Both agencies (NIH, FDA), influenced by the OSTP mandate to foster biotechnology and avoid restrictive regulations, and faced with considerable uncertainty about risks posed by new genetic materials and inability to discern such risks before human experimentation, chose to rely on managing patient safety by means of organizational learning from experimental experience.

Thus, they established procedures for clinical trials and other experimentation with humans in which researchers are expected to follow a precautionary protocol for selecting patients and applying experimental methods and products to them, monitoring their response, and then do full and timely reporting of any adverse outcomes to enable open discussion and evaluation of the causes, quick determination of corrective actions needed for patient safety, and rapid dissemination and implementation of the corrective actions in all related research projects. It now appears that this model for safeguarding human subjects needs reinforcement for it to be sufficient in the competitive and economically opportunistic culture in which gene therapies are tested.

The gene therapy experience also implicates the latent social control of legal liability for medical malpractice. Whereas routine medical practice is readily held accountable for errors and others injurious renderings of inferior care to patients by litigation in which medical malpractice liability doctrines are applied, clinical testing is experimental, is done on patients willing to take therapeutic risks because of their desperate condition, and poses great difficulties for proving causation. As a result, the application of malpractice law and imposition of liability on medical researchers is far more problematic. Thus, malpractice liability is a relatively weak form of social control for ensuring patient safety in the clinical testing of new genetic therapies and products (Palmer, 1999).

Reliance on yet another social control to secure patient safety in gene therapy trials, self-regulation by medical professionals, also seems misplaced. Although medical researchers are licensed doctors subject to professional codes which firmly establish their responsibilities to heal patients and "do no harm," they are also researchers who view persons in clinical trials as human subjects, certainly deserving of care, but nevertheless as willing subjects of experimentation with their safety, privacy, and other interests subsumed to the larger purposes of research.

Thus, unless malpractice liability is made more clearly applicable to clinical trials and medical researchers are held to the responsibilities of treating physicians, the burden of social control rests on improved forms of agency regulation. The need for more effective regulation to prevent harms is now being deliberated and is expected to lead to increased regulatory requirements for organizational responsibility and more effective oversight and control of experimental work by organizational review boards. A new watchdog agency to protect patients in medical research, the Office of Human Research Protection, is expected to be created to oversee private and federally funded trials. If proposed legislation is enacted, the new agency will be given authority to impose penalties on researchers of up to $250,000, and on research organizations of up to $1 million, for violating agency requirements for studies involving humans (National Institutes of Health, 1999; Stolberg, 2000).

Also expected are new rules requiring that researchers disclose their financial interests to the patients involved, secure renewed consent from the patients if serious problems or deaths occur in such trials, comply with more definitive reporting requirements for adverse events, implement better monitoring, and attend training programs on ethical research and patient safety. Research organizations will also be responsible for assuring these requirements are met, and will be required to conduct compliance audits. Organizations wishing to continue with clinical testing will undoubtedly have to invest in and install safety management systems that are far more complex and vigilant than heretofore, a major challenge since most of the organizations are academic or medical institutions which have not developed sophisticated safety management systems of the types employed in the nuclear and chemical industries.

Anticipating these burdens, a major medical research center (at the University of Pennsylvania) has decided to outsource any human testing to other organizations and confine its on-site genetic research to molecular, cellular and animal studies. Such outsourcing and other forms of organizational change pose additional challenges for safety management systems (Hale and Baram, 1998). Critics of these reforms, because of the burdens, include academic and medical administrators and researchers. However, specialists in research ethics believe the reforms are inadequate and primarily intended to avoid interrupting clinical tests of pharmaceuticals, gene therapies and other applied research with high commercial potential.

c. Agricultural sector

As biotechnology has advanced over the last decade, numerous genetically modified crops, bacterial strains, and livestock species have been developed with corporate and government funding. Federal agencies, following the OSTP policy, have established relatively permissive regulations for field testing of these genetically created products, and approved many for commercial distribution, sale, and use in the agricultural sector. The U.S. Department of Agriculture (DA) and the Environmental Protection Agency (EPA), with authority to protect agriculture and the environment, have also waived testing requirements for many modified crops in recent years, further expediting quick entry of such products into commercial markets. The agencies have justified the recent waivers on the lack of evidence of risk in thousands of field tests done for earlier modifications of such crops. To scientific and lay critics, the agency rationale for waivers is flawed because the early field tests were limited and the genetic modifications involved differ from the more recent genetic modifications. Similarly, the Food and Drug Administration (FDA), authorized to ensure food safety, has refrained from setting requisites for pre-market testing or protective standards, and resisted strong pressures from consumer groups that foods with genetically-modified constituents be labeled as such (Committee on Science, 2000).

Genetically-modified crops (e.g., cotton, soybeans, wheat, corn, rapeseed, potatoes, squash, etc.) are now aggressively marketed by the small number of American and European firms that lead in this field (e.g., Novartis, DuPont, Monsanto, Zeneca, etc.) in the U.S., Europe and elsewhere as being more resistant to pests, having longer "shelf-life", providing superior nutrition, and posing other benefits including greater productivity and profit potential for farmers, cheaper and healthier foods for consumers, and more reliable food supply for poorer societies ravaged by periodic famines. These marketing initiatives have been most strongly opposed by European consumers, and more recently, by a growing number of Americans, and have also met with resistance in a number of other countries (Baram, 1996; Paarlberg, 2000).

Opposition, initially based on unproven fears of latent risks to the environment and to consumer health, as well as on the aggressive tactics used by such multinational chemical companies to move into and dominate food production systems, has intensified as evidence of risks has accumulated in recent years. For example, Danish researchers have produced findings that the new genes added to such crops can and do flow to related plants in the natural environment; British researchers have demonstrated that genetically-modified potatoes harm the health of laboratory test animals; Brazilian health officials have confirmed that consumers allergic to nuts have been harmed by unlabeled foods containing nut genes; and American researchers have found that new pest resistant corn interferes with the life cycle of the Monarch butterfly (Bureau of National Affairs, 2000; Baram 2000).

Opponents have also developed the hypothesis that crops genetically-designed to kill targeted species of pests, being far more effective than chemical pesticides, essentially restrict mating to highly-resistant survivor pests and thereby accelerates the continual creation of ever more resistant pest species. Thus, they contend that such modified crops accelerate the rate at which resistant pests are created and the level of resistance. Although agencies and industry initially resisted this hypothesis, further study has led them to accept it. However, it remains the fact that remedial action is left to the discretion of the companies and to farm practices (e.g., use of computer-designed measures for alternating plantings and delineation of buffer zones between conventional and genetically modified crop areas, etc.) and is therefore uncertain (Levidow, 1999).
Because more stringent risk regulation (e.g., more extensive testing before agency approval for sale, labeling, etc.) has been opposed by American agencies and, the multinational companies, market resistance in Europe and the U.S. has grown. This has led farmers, food processing companies, and food purveyors to increasingly avoid genetically designed crops, and to favor more costly conventional crops and food products. In addition, resistance to labeling, a necessary measure for protecting consumers with allergies and for enabling other consumers to practice dietary rules for health, religious, or other reasons, has outraged many consumers and further diminished the profit potential of genetically-modified products in agriculture (Baram 2000).

As a result, market forces are causing U.S. agencies and multinational companies to temper their resistance to reform and changes are anticipated. Several agency and company initiatives for accepting some form of labeling are now being developed, and test requirements are being re-evaluated to make them more credible. Thus, although regulation has thus far failed as a means of social control for ensuring public health and environmental protection, it has become clear that the most unpredictable form of social control, the exercise of popular choice in the marketplace, has been activated and is likely to exert continuing pressure for improved regulatory safeguards in bioagriculture (Hoban, 2000).

4. CONCLUSION

This brief survey of the advance of biotechnology and the spontaneous emergence and evolution of disparate social controls, their interaction and early outcomes, indicates the complexities of adapting the most powerful new technology to socio-economic systems and cultures. Ad hoc responses to moral and ethical implications by national officials, and incremental improvements in risk regulation, arising from research findings, adverse events and their influence on public perception and market behavior, are the dominant features of how the U.S. is coping with biotechnological advance.

Traditional approaches to managing the safety of new technologies, which emphasize pro-active regulation and the installation of elaborate safety management systems in agencies and industrial organizations, have not been applied to biotechnology in the health care and agricultural sectors. However, recent initiatives have been driven by growing public pressures, and further reforms are imminent. These ad hoc developments indicate that coherent safety management systems are needed. Thus, expertise in safety management, developed for other technologies, is likely to become a valuable resource for developing more effective social controls for biotechnology.

REFERENCES

Baram, M. (1996). LMO's: Treasure Chest or Pandora's Box. *Environmental Health Perspectives*, v. 104, n. 7 (1996) pp. 704-707.

Baram, M. (1997). The Laws of Genetics. *Environmental Health Perspectives*, v. 105, n. 5, p. 488.

Baram, M. (1999). Biotech Battles. *ChemicalWeek*, (Dec. 15, 1999) pp. 22-24

Bureau of National Affairs (Feb. 2 2000). *International Environment Reporter,* pp. 90-92.

Bush, D. (2000). The Role of the National Institute of Health and Conflicts of Interest. *Biotechnology Law Report*, v. 19, n. 5 (Oct. 2000) 576.

Caulfield, T., B. Williams-Jones (1999). *The Commercialization of Genetic Research.* Kluwer/Plenum.

Committee on Science, U.S. Congress (April 13, 2000). *Seeds of Opportunity: An Assessment of the Benefits, Safety and Oversight of Plant Genomics and Agricultural Biotechnology.*

DeLisi, Dr. Charles (Boston University Bioinformatics Program). Personal communications.

Franz , N. (2000). DoE Increases Funding for Bio Feedstocks R & D. *Chemical Week*, March 22, p.48.

Green, A. (2000). Increased Industry- Academic Interactions Lead FDA to Require Financial Disclosure by Clinical Investigators. *Journal of BioLaw & Business*, v.3, n.2 (2000) pp.15-20.

Grubb, P. (1999). *Patents for Chemical, Pharmaceuticals and Biotechnology.* Clarendon Press.

Hale, A., M. Baram (1998). *Safety Management: The Challenge of Change.* Pergamon.

Heller, M. et al (1998). Can Patents Deter Innovation?: The Anticommons in Biomedical Research. *Science,* v. 280, p.698.

Hoban, T. (2000). Social Controvery and Consumer Acceptance of Agricultural Biotechnology. *Journal of BioLaw & Business*, v. 3, n. 3, (2000) 38-46.

Human Genome News (nov. 2000). v. 11, n. 1-2.

Icelandic Health Care Database- deCode Genetics. Inc. website: http://www.database.is/almennt.html>.

Kowalczyk, L. (2000). Group hits Hub Doctor facing FDA Scrutiny: Possible Financial Conflict is Seen, *Boston Globe* (May 13, 2000) p. C-1.

Levidow, L. (1999). Regulating Bt Maize in the United States and Europe, and Insect Resistance. *Environment*, v. 41, n. 10 (Dec. 1999) pp.10-22.

Merges, R. et al (1997). *Intellectual Property in the New Technological Age*. Aspen.

National Institutes of Health (Nov. 16, 1999). Recombinant DNA Research: Proposed Actions Under The NIH Guidelines. *Biotechnology Law Report*, v. 19, n. 2 (April 2000) pp.204-205.

Office of Science and Technology Policy (1986). Coordinated Framework for Regulation of Biotechnology. *Federal Register*, v. 51, n. 123, June 26. pp. 23302-23309.

Office of Science and Technology Policy (1992). Exercise of Federal Oversight Within Scope of Statutory Authority: Planned Introductions of Biotechnology Products into the Environment. *Federal Register*, v. 57, n. 39, Feb. 27. pp. 6753-6762.

Paarlberg, R. (2000). *Environment*, v. 42, n. 1 (Jan. 2000) pp. 19-27.

Palmer, L. (1999). Patient Safety, Risk Reduction, and the Law. *Houston Law Review*, v.36 (1999) pp.1609-1661.

Rifkin, J. (1998). *The Biotech Century*. Penguin Putnam.

Secretary's Advisory Committee on Genetic Testing, National Institutes of Health. (12 april 2000). Adequacy of Oversight of Genetic Tests and other Committee reports at their website: <http://www.nih.gov/oba/sacgt.htm>

Sharp, R., J.C. Barret (1999). The Environmental Genome Project and Bioethics. *Kennedy Institute of Ethics Journal*, v. 9, n. 2.

Stolberg, S. (2000). Fines Proposed for Violations of Human Research Rules. *NY Times* (May 24, 2000) p. A-21.

Testimony of Amy Patterson, M. D. (2000). NIH-Office of Biotechnology Activities. *Biotechnology Law Report*, v. 19, n. 2 (April 2000) pp.192-201.

Vincent, C., B. de Mol (2000). *Safety in Medicine*. Pergamon.

The final paper in this section discusses another regulatory frontier - the rise of certification by private sector organisations. This is not just a matter of self-regulation, because it involves certification bodies carrying out audits and inspections of other bodies, in both the public and private sectors, in ways which are quite similar to the new auditing functions of government regulators. What is driving this development is that many businesses will now only do business with others whose products, services or processes have been certified by some third party as complying with relevant standards.

Gundlach describes these developments in considerable detail. He notes that certification bodies are at present too narrowly constituted and need to incorporate consumer and trade union elements to strengthen them. He notes further that certification bodies can experience severe conflicts of interest and that these should be more effectively banned.

Gundlach argues that private sector certification has become an important adjunct to government regulation. He raises the interesting question of whether it might replace government regulation altogether and he answers this with a clear 'no'. Voluntary third party certifiers cannot enforce compliance and there will always be a need for government agencies to carry out this function.

Gundlach is clearly an advocate of certification as a tool, but we still have few studies of its cost-effectiveness, or whether its pitfalls are really likely to be significantly less than those of the central government regulation it so often replaces. Unarguably it shifts many of the costs of regulator to the regulated, rather than to the general society via the tax revenue.

14

CERTIFICATION, A TOOL FOR SAFETY REGULATION?

Harry Gundlach

INTRODUCTION

Regulatory bodies are increasingly using accreditation and (third party) certification as an impartial independent and transparent means of assessing that organisations are abiding by laws. Governments are being forced to use (private) accredited certification instead of governmental inspection and supervision by rapid and dynamic change in technologies and the shortage in manpower.

In this paper third party certification and accreditation are described, including the regulatory technique and strategy used in the 'New and Global Approach'. Questions concerning the applicability of certification in (safety) regulation in the Dutch situation are raised, particularly about the elimination of parliamentary influence in cases of interpreting laws, the non-representation of consumers and trade unions in 'committees of experts' and last but not least the lack of the power of enforcement. The involvement and commitment of governments in the boards of accreditation bodies, which control certification, is another question mark. Finally the kind of certification applicable to health and safety regulation is discussed.

In the past regulators functioned largely as inspectors, operating in the field and checking whether companies were in compliance with detailed legal requirements. Nowadays the focus is on management systems and inspectorates have to assess the implementation of these systems. A very different set of skills is needed and many inspectorates have difficulties retooling and retraining to deal with this new way of working. The rapid and dynamic change in technologies and the shortage in manpower is forcing governments to use private (accredited) certification instead of governmental inspection and supervision. Therefore more and more government authorities look to (third party) certification as a reliable tool in helping them discharge their work functions in protecting the safety of people, property and the environment.

It requires professional judgement to determine the suitability of safety management systems and safety aspects of products, for example, in securing compliance with legislation. Parties involved will have to spend much time and money to find and train personnel with the necessary competence to perform these tasks. For inspectorates however the move to self-regulation is often accompanied by a reduction in available resources. The cost of specially trained personnel and expensive equipment are too much of an expense today to be passed on to the taxpayer.

CERTIFICATION

Certification of products goes back hundreds of years. Early examples are the acceptance of craft guilds for building castles, palaces and churches in medieval times, for the proofing of gun barrels, for the assay and hallmarking of metals like gold and silver, and for the certification of both guild brothers and their products in the 16th and 17th centuries (Gundlach, 1988). Marks on products were and are seen as information carriers, a short hand means of providing product information for consumers and professional purchasers.

The modern definition of (third party) certification reads: "Certification is a procedure by which a third party gives written assurance that a product, process or service conforms to specific requirements", (ISO/IEC Guide 8402, 1994). According to the definition, the kinds of certification are:

- certification of goods and services, generally named certification of products,
- certification of processes, e.g. the process of wood preservation,
- certification of personnel, assuring the professional skill of personnel. and,
- certification of systems (e.g. quality management system based on ISO 9000).

Certification can be used by suppliers as a means of production control as well as a contribution to a better operation of the design and manufacturing process. Certification to ISO 9000 standards has been the fastest growing certification practice of all times, and is increasingly seen as a 'ticket' for entry into important markets. Distinct from mandatory systems, the success of private certification is caused by the voluntary participation of the parties involved.

Certification can be split into four activities (ISO/IEC, 1992):

- Initial assessment of the quality system of the manufacturer and or testing of the product;
- Appraisal and weighing of the results of the assessments and test reports;
- Decision on certification which can result in a 'Certificate of Approval', and
- Certification surveillance visits.

Certification Surveillance Visits

In addition to the initial assessment, market surveillance also belongs to the core business of certification. Surveillance is a kind of market supervision and or monitoring, undertaken by the certification body and should give assurance that certified products and or (quality management) systems continue to comply with the criteria. Typical surveillance activities are testing and inspection and/or assessment of the quality system of the supplier. Samples for surveillance testing and inspection should be typical of production and preferably selected from the factory and/or the open market. Visits by inspection and/or audit personnel of the certification body to the factory/supplier where certified products are produced frequently involves the assessment of (elements of) the quality management system based on the ISO 9001 standard. Certification surveillance visits should be distinguished from the surveillance visits by or on behalf of public authorities to enforce suppliers to meet legal requirements.

Validity of a Certificate

A certificate (of Approval) is normally valid for a period of 3 to 4 years after which a reassessment takes place. At any time within the validity of the certificate the certification body retains the right to suspend or withdraw the certificate and or cancel the contract. The certification body may enforce these actions if the supplier is operating in contravention of the requirements. This includes misuse in any way of the mark of conformity or operations that could bring the certification process into disrepute.

If during a routine surveillance a major non-conformity is discovered the supplier is given a fixed time to put it right. Suspension occurs normally if an extra surveillance visit, which is conducted after a routine surveillance, finds serious non-conformities, indicating that inadequate or no corrective actions have been taken by the supplier to clear the non-conformities uncovered previously. If the supplier has been unwilling or unable to clear the non-conformities within reasonable time, the certificate may be withdrawn.

Withdrawal of a Certificate is a difficult and serious step, which is initiated only when it becomes apparent that corrective actions will be unsuccessful in bringing about full compliance with the requirements. The supplier may appeal against all the decisions of the certification body, including the withdrawal of the certificate. On the other hand, because certification bodies are earning money with the certification activities and due to the competition in the certification market, certification bodies may be reluctant to withdraw a certificate for non-conformity.

WITHDRAWAL ISO 9000 CERTIFICATES IN 1999	
• not meeting requirements	643
• moved to another Certification Body	676
• Termination of business	884
• Not enough adde value of certificate	473
• Unknown reasons	7186
Total	9862

figure 1. Withdrawal of certificates

In 1999 world wide 9862 ISO 9000 and 83 ISO 14000 certificates were withdrawn for the reasons shown in figure 1 (ISO ISO9000 and ISO1400 survey). At that time a total of 340,000 certificates were current. From these figures and from information delivered by some of the largest certification bodies can be concluded that round about 0.5%- 1% of the certificates are withdrawn by them due to fact that manufacturers/suppliers were not meeting the standards.

ACCREDITATION

Purchasers and consumers ask for a structured and transparent conformity assessment process in which impartiality and competency is crucial for their trust. Accreditation can provide impartial and independent evidence of the ability of conformity assessment organisations e.g. certification bodies to provide satisfactory certification. Conformity declarations, which are issued under accreditation, express the confidence that organisations work in accordance with a pre-determined standard so that the infrastructure is demonstrably present for supplying high-quality services and products.

Accreditation is by definition a procedure in which an authoritative body gives formal recognition that a body or person is competent to carry out specific tasks. An accreditation body evaluates the competence of a conformity assessment body, like certification, testing and inspection bodies, on the basis of internationally agreed criteria. These criteria contain requirements concerning impartiality, participation of all parties concerned and competence of the assessment personnel. For product certification e.g. the following 'hierarchy' exists:

- Accreditation body that recognises the competence of a third-party certification body for a sector of products.
- Certification body (third-party) which declares conformance of products with e.g. standards or other normative documents and licences their mark or logo.
- Supplier/manufacturer (first-party) placing products with the 'third party' mark on the market.
- Purchasers, users and/or consumers (second-party), who buy/use the products with marks.

Accredited certificates fulfil the role of a reliable 'information carrier' (Graaf de, 1996) and can inform governments that a company operates a management system which conforms to the applicable standard, which could include occupational and health items.

Accreditation is valid for a limited time and applies to scopes of products and or management systems. Accreditation is normally valid for a period of 3 to 4 years after which a reassessment takes place. The accreditation body retains the right to suspend or withdraw the accreditation and or cancel the contract. However in contradistinction to certification, withdrawal of accreditation of a certification body concerns more parties including all the certificate holders. In the Netherlands the Dutch Council for Accreditation since its foundation in 1982 has withdrawn only a very few accreditations.

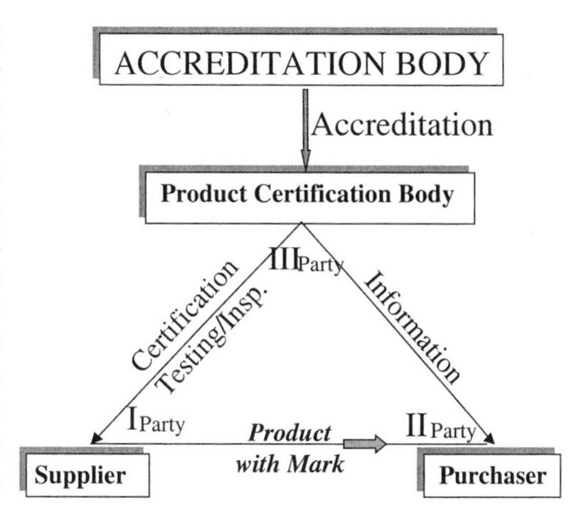

figure 2. Structure of Accreditation

NEW AND GLOBAL APPROACH

An extra impulse to the use of accreditation and certification since 1985 is the so-called 'New Approach' to product regulation and especially the 'Global Approach' to certification and testing. The aim of the 'New Approach' was and is to remove technical barriers to trade (caused by health and safety regulation) and to stimulate free movement of products within the European Community. The 'Global Approach' lays down general guidelines and detailed procedures for conformity assessment such as certification, testing and inspection that are to be used in European directives (European Commission, 1999)

European directives are an important instrument of the 'New Approach' in the field of technical harmonisation and standardisation. They are intended to remove the technical barriers by harmonising product regulations and conformity evaluations. No specifications but only 'essential requirements' restricted to the protection of health and safety of users and the protection of environment, are laid down in the New Approach directives. Up to 2000 some 20 directives have come into force. The directives cover products where safety aspects are relevant such as lifts, pressure equipment, explosion-proof materials, toys and personal safety equipment.

NOTIFIED BODY

'New Approach' directives prescribe, besides the 'essential requirements', when it is that the manufacturer during the design and/or production stage is required to have a 'Notified Body' carry out conformity evaluations of their product and/or quality management system. To gain a recognised/registered status as 'Notified Body' a certification testing or inspection body must be designated and notified by an EU Member State. Forr the purposes of designation the Member State must ensure that the body satisfies at least the minimum designation criteria specified in the directives. Demonstrating congruity with the harmonised European standards in the EN-45000 series can usually provide proof of this (figure 3). The EU-Council of Ministers has agreed that the ministers involved at a national level will make as much use as possible of accreditation to demonstrate this congruity.

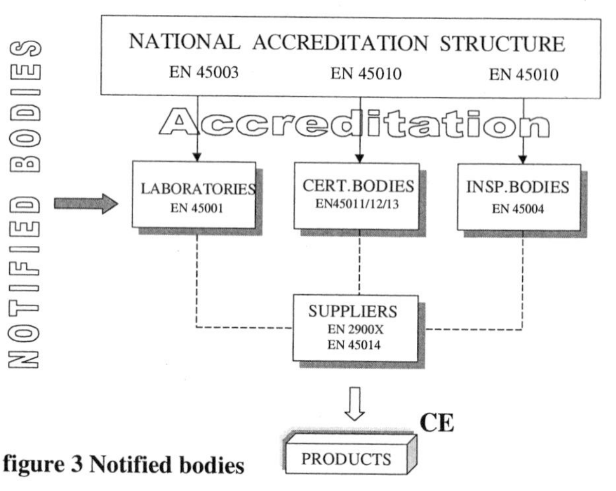

figure 3 Notified bodies

The description of the certification system up to this point in the chapter has been that applicable in the whole European Union. Many aspects apply also outside the EU. In what follows the working of the system in the Netherlands forms the focus, though many of the concerns and points made do have much wider relevance and application.

Participation of All Parties Concerned

In specifying the criteria against which the products or the quality system of an organisation are evaluated and how they should be assessed, a certification body has to establish working documents. The working documents are part of a certification scheme that reflects the content and functioning of the certification. Inherent to certification is a structure that should enable the participation of all relevant parties concerned to approve a certification scheme. Representation on an Advisory Board and or in an impartial committee of experts forms the means to take into account the views of the significant interested parties.

An interesting case related to this point of participation of all relevant parties in approving a certification scheme, was one in The Netherlands in 1997. A certification body was challenged by some consumers for licensing its mark to be used on wood preserved using copper-tungsten additives. At that time the copper-tungsten additive was allowed by law, but was regarded by 'environmental organisations' as unsafe and environmentally dangerous. One of the arguments of the consumers was that the member's chosen in the board and in the 'Committee of technical experts' of the certifying body were not balanced and that the single interest of manufacturers predominated. The defence of the certification body was that the certification scheme was in conformity with the law and that they were accredited for this certification scheme. Complainers however argued that it makes no sense to use voluntary certification just to assess conformity with the law, because this is an obligation for everyone anyway. Voluntary certification of wood should only be used if it has added value, e.g. if the certification protocol exceeded the requirements of the law by banning copper-tungsten as a preservative. In their view a more balanced participation of all parties concerned would have resulted in an environmentally friendly and more realistic certification scheme.

It is essential for the status of voluntary certification that all identifiable major interests should be given the opportunity to participate, and that a balance of interests is achieved, where no single interest predominates. However for accreditation it was and is difficult to assess the balance in representation especially if consumer and environmental organisations are at stake, due to the number of organisations concerned and their different views.

Interpretation of Laws

Impartial committees, the so-called 'Committees of technical experts' (CoE) are persons, possessing the necessary technical competence to, formulate explanations and interpretations where required for the application of the criteria for a specific certification scheme. Where this is done by such committees of the main parties concerned, the interpretation of the law is 'shifting' from the authorities to the certification bodies. In principle the work of these committees should be restricted to giving guidance on the understanding of the criteria. However the boundaries of the term "guidance" are difficult to define and in practice it sometimes shades into giving technical specifications as well.

Where requirements and especially technical specifications are laid down in legislation a problem exists concerning the legal repercussions of the explanations and interpretations proposed by the CoE and by the Advisory Board. In the past many inspectorates had representatives on these CoE. In this way authorities could influence the interpretations and bring them in line with the intention of the law. At the same time they could acquire and keep up their competence and practical experience of the regulated activity. At the present time, due amongst other things to the cutting of inspectorate manpower, the increased number of certification bodies, and a change in government policy, there is practically no government involvement in these committees. Parliament is also not involved, despite the fact that it is responsible for establishing the legislation. Hence, the interpretation or "tuning" given to the regulations in the translation process into certification protocols by all other parties concerned can differ from the intention of parliament, the government and its inspectorates.

A Better Regulatory Technique

A new regulatory technique and strategy has been introduced by the 'New Approach' to technical harmonisation and standardisation. The 'New Approach' focuses on the performance of products, relating to the (local) market needs.

The following principles have been established:

- Legislative harmonisation in Directives is limited to essential requirements that products placed on the market must meet, focused on the protection of health, safety and the environment.
- Products meeting the technical specifications laid down in harmonised EN-standards, issued by the European Standardisation Body CEN/CENELEC, have the presumption of conformity to the essential requirements set out in the directives.
- Application of harmonised standards remains voluntary, and the manufacturer may always apply other technical specifications to meet the requirements. The manufacturer/supplier then has the obligation to prove that his products are in conformity with the essential requirements using his own specifications. The manufacturer/supplier may be obliged, dependent on the

requirements in the applicable directive, to submit this proof of conformance to the assessment of a third party (notified body).

Essential requirements must be applied as a function of the hazards and risk level inherent in a given product. Therefore suppliers need to carry out a risk analysis to determine the essential requirements applicable to their products. Products may be placed on the market and used only if they are in compliance with the essential requirements and, as proof thereof, bear the CE-marking.

How to Involve All Parties

The principle that technical specifications are or have to be laid down in harmonised EN-standards introduces the process of drafting (EN-) standards. In principle the new regulatory technique takes the line that all relevant parties should be consulted in separating requirements into what is essential and what is not. For aspects of safety and protection of health the input of consumers and trade unions is essential. However, it is becoming more and more difficult to take into account the views of these significant interested parties. As a consequence of a lack of funding and manpower, these parties are not in the position to join all the committees. To solve these problems more creative approaches need to be considered like telephone conferencing and/or the use of web sites. For consumer organisations representation by retired members is an option to solve the lack of manpower. In the Netherlands 'central committees of experts' have been established for certification and notification in some sectors to reduce the load of committee attendance otherwise needed if each certifying body were to have its own CoE.

Central Committees of Experts in the Netherlands

In the field of occupational health and safety (ARBO in Dutch) and under the terms of the 'Dangerous Equipment Act' the Ministry of Social Affairs and Employment is responsible for the implementation of a number of directives. It is encouraging the use of the accreditation and certification instruments under the auspices of the Dutch Council for Accreditation. A certification/inspection or testing body which wants to operate as a Notified Body for a specific field of work covered by one of the European directives, has to apply for accreditation (Bennink, 1998). Normally the bases for accreditation are the criteria in EN 45011 (for product approval) or in EN 45004 (for inspection work). However these criteria do not cover all the requirements that the Notified Bodies has to meet according to current legislation and regulations. The Ministry reached an agreement with the Dutch Council for Accreditation for the drafting of so-called 'Directive specific accreditation schemes' (RISAS's) covering additional requirements. The RISAS's have been drawn up by all parties concerned in Central Committees of Experts on the basis of the relevant elements taken from both EN-standards, supplemented with regulatory criteria.

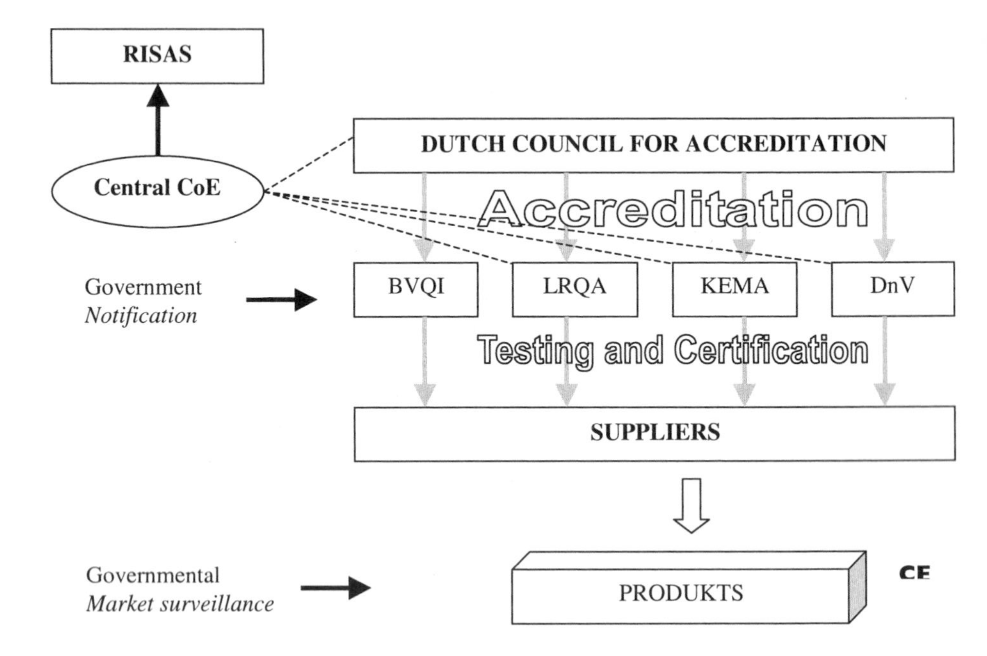

figure 4. Public Sector – private sector relationships

The RISAS embodies the interpretation of the (regulatory) criteria which certification and inspection bodies are required to satisfy. The RISAS also includes an overall delineation of the procedures prescribed in the Directive. This avoids inconsistencies arising between the designated Notified Bodies when procedures are implemented. RISAS's do not infringe the individual procedures and instructions of the concerning body and do not contain any additional technical requirements (Bennink, 1998).

Accreditation on the basis of a RISAS will be compelling evidence for the Minister of Social Affairs and Employment in designating Certifying Institutions as approval bodies, and having these subsequently notified as Notified Bodies. For the time being, if no RISAS is in place, Notified Bodies are given a temporary designation from the Minister of Social Affairs and Employment. The following RISAS's have been developed or are in draft to date:

- Simple pressure vessels,

- Pressure equipment,
- Machine directive.
- Personal safety equipment.
- Lifts.

The Ministry of Social Affairs and Employment intends to submit the various directive-specific accreditation plans (RISAS's) for notification in Brussels.

Voluntary Use of Harmonised Standards

As stated above, another principle of the 'New Approach' is that application of harmonised standards is voluntary. Harmonised standards with detailed technical specifications can act as a brake on the advancement of technology – the state of the art -because the technical specifications of products need to and are changed frequently as technology advances. If a detailed standard must be changed, this takes time. Under the New Approach, the manufacturer/supplier has the obligation to prove that his products are in conformity with the essential requirements. He can use instead of harmonised standards other means of his own choice, for example by means of any existing technical specifications. The essential requirements do not change, only the manufacturer/supplier's dossier which demonstrates compliance, a much faster process. However 'existing technical specifications' in contrast to 'harmonised standards' are in many cases drafted only by technical experts and miss the contribution of other parties involved, like users.

Supervision and Enforcement

Research by consumer organisations in Europe (and USA) and the results of the market surveillance authorities have shown that a substantial number of products are not safe, notwithstanding the 'New Approach' and the efforts of industry to convince the consumers that industry acts in a responsible way. Currently recalls of products are relatively commonplace. This is partly due to inadequacies in the standards, but in most cases it is caused by failures in the operation of the manufacturing process (Gundlach, 1999).

Market surveillance is an essential tool for enforcing New Approach directives by taking measures to check that products remain safe, and meet the essential requirements. Member States have to nominate authorities, with the necessary resources and powers, to be responsible for market surveillance. The question is if and how Governments can use private third-party certification in performing market surveillance. Apart from the initial assessment, market surveillance belongs also to the core business of certification. However distinction should be made between:

- supervision of voluntary certification (certification surveillance visits) outside legal frameworks, and
- supervision and enforcement of legal requirements.

The latter principally belongs to the power of the public authorities only. Art. 10 of the EC Treaty explicitly requires Member States to take all appropriate measures to ensure fulfilment of their obligation to enforce Community legislation (European Commission, 1999). The national surveillance authorities may subcontract tasks to e.g. certification bodies, provided that they remain responsible for their decisions. Enforcement can result in penalties, including the obligation to pay fines or even to close factories. Certification bodies doing market surveillance for national authorities are not allowed to do other conformity activities that raise conflicts of interest.

Private certification is focused on satisfying market needs. Like most other organisations, certification bodies operate a competitive business and have contractual agreements with their clients to the effect that the supplier will market only products conforming to the standard. Market surveillance is part of the contractual agreement and belongs to the supervision task of the certification body. If non-compliances are discovered the certification body has to oblige the supplier to make corrective actions and can even go for prosecution under civil law. However no legal enforcement is possible. At the end of the day the certification body can only withdraw or suspend its certification.

It can happen that a certification body performing market surveillance detects unsafe products. In principle this is a violation of the law, which should be reported to the relevant government inspection agency. However, confidentiality is an important aspect of certification. This is a conflict of interest that has to be solved. It can also limit in general the use of certification by regulatory bodies as a means of assessing that organisations are abiding by laws. There is a need at least for a code of practice and/or of conduct for certification bodies finding non-conformities to the law.

IMPARTIALITY AND INDEPENDENCE OF CERTIFICATION

In the context of many commercial activities among participants in the economy, and in the context of mandatory regulations, the added value of conformity assessment is the level of confidence that the economic partners have that the subject of certification meets their expectations. In this context the value of certification is dependent on the perceived integrity of the accreditation and certification structures involved.

Discussions are currently going on about what independence, impartiality, integrity and potential and actual conflicts of interest are and under what circumstances they might influence the integrity of the certification process. Impartiality and independence of the certification body should be assured at three levels:

- Strategy and Policy;
- Decisions on Certification, and
- Evaluation.

Independence designates a quality or state of not being subject to the rule or control of another i.e. being self-governing. It stresses the state of standing alone not depending on others for one's opinion or livelihood. Impartiality means freedom from favour or prejudice toward the one or other side. Integrity can be described as the state of mind of the staff while independence and impartiality are attributes of the certification body concerned.

For authorities, purchasers and consumers objectivity and impartiality are crucial for their trust in certification. Conflicts of interest of the certification process and structure can be split up into two categories: inherent and occasional or structural conflicts.

Inherent conflicts of interest

- Caused by the fact that a certification body receives money for the certificates issued;

The manufacturer may be only interested in 'paper on the wall' instead of in implementing or maintaining a quality system. In this case he will most likely choose for the cheapest and easiest way to get the certificate, irrespective of the quality level.

- Caused by the audit process itself;

When does an auditor cross the border between making a judgement about conformity, where it does not meet the audit criteria, and becoming a consultant telling the company what to do to conform?

Structural conflicts of interest

- Caused by a combination of adjacent business;

Accreditation rules do not allow the Certification Body also to provide consulting services to obtain or maintain certification/registration within the same organisational entity because of the conflicts of interest that can arise. Consultancy is considered to be participation in an active creative manner in the development of the item (e.g. management system) to be assessed. If the certification body offers accredited certification to those organisations to which it has provided consulting, it cannot be objective in its evaluation of the system, since it has helped to develop it. However certification bodies are allowed to carry out some activities without them being considered as consultancy; e.g.

- Training courses related to certification. However, these should confine themselves to the provision of generic information and advice, which is freely available in the public domain. So open courses are acceptable, but company specific training is not allowed;
- Identifying opportunities for improvement which become evident during the audit or assessment, provided that they do not recommend specific solutions.

- Caused by a structure in the certification body with an unbalanced representation of stakeholders;

Sometimes board or committee members are chosen from related bodies, which are linked to the certification body by common ownership, contractual arrangement or informal understanding, i.e. which have too close a relationship with the certifying body. The board then does not provide a balance of interests where no single interest predominates.

Certification bodies have recognised the danger of combining consultancy and certification services. The Guideline for auditing Quality systems (ISO 10011-1 future 17011) as well as the accreditation standard for Certification bodies (ISO62/65/EN45012/11) discourage – not to say prohibit - that Certification bodies and their auditors become involved in 'consultancy', and in doing so lose their impartiality and objectivity. However, many smart structures have been invented which cause much confusion in the market. For example, besides holding companies, which contain both certification bodies and consultancy agencies, less transparent structures such as 'virtual' networks are becoming active.

People who have been involved, in the last two years, in activities such as consultancy for a supplier asking for certification, cannot be employed to conduct an evaluation as part of the certification process. Confusing to the market is that the same person is often contracted for both audit and consultant functions and has business cards on which both functions are mentioned.

GOVERNMENTS IN BOARDS OF ACCREDITATION

Accreditation is used by Regulatory Bodies as an impartial and transparent means of assessing the competence of conformity assessment bodies like laboratories, certification bodies and inspection bodies. In some countries accreditation is in private hands, however with strong links with governments through, for example, representation in their Governing Boards. Impartiality can, in conformance with the ISO/IEC Guide 61, only be safeguarded by a structure that enables the participation of all stakeholders in the development and principles governing the content and functioning of accreditation. The participation of all stakeholders in the Board of Governors has or should have the effect of counteracting any tendency of the government, under whose auspices an accreditation body operates, to exert undue influence on the decisions on accreditation.

However, it is useful to subject this government participation to the test of the principles developed by Montesquieu known as the 'Trias Politica'. These aim to separate the functions of government into three distinct branches: the legislature for making the laws; the executive branch for the enforcement of those laws; and the judiciary branch for the interpretation and scrutiny of the law. By having such a system in place, governmental power is separated and balanced in a way which guarantees individual rights and freedom.

Accreditation can be seen as a judiciary power. Therefore governments (i.e. the legislature or the executive) which use accreditation should not be directly involved in the governing bodies concerned with it. Most governments do use accreditation to notify the conformity assessment bodies. These notified bodies carry out the task of checking conformity to the relevant directives, when a third party is required. Notified bodies are active in areas of public interests and, therefore, are answerable to the public authorities. Licensing fits into the same category as notification, because here too conformity assessments are focused on regulated items. The conflict of interests between the three arms of the Trias Politica is even more critical when governments are carrying out certification, testing and inspection themselves.

In all of these cases it is crucial that government should not have a say, directly or indirectly, in the process of accreditation of all these certification and assessment bodies. It should not sit in the governing boards, but instead, should be represented in the 'Committees of Experts' as a normal user of accredited certification.

System Certification

The chapter now turns to the area of system certification, notably of management systems and looks at its applicability to the area of health and safety.

Quality Management System (QMS)

'Allied Quality Assurance Procedures' (AQAP) standards have been used throughout NATO for all defence related suppliers since the 1960s. Based on this experience, the benefits of QMS procedures became apparent to all stakeholders including consumers and professional purchasers. The need for QMS-standards, which could be used by industry in general, was recognised. In many countries local QMS-standards were developed. In the UK this led to the development of the pioneering British Standard BS 5750 in 1979. This formed the basis of the ISO 9000 quality management standards developed by the International Standardisation Organisation (ISO/IEC) at Geneva in 1987. The ISO 9000 series of International Standards for quality management is among the most widely known and successful of the 13 000 standards published by ISO since it began operations in 1947.

The ISO 9000 standards have become an international reference for quality requirements in business to business dealings and form the basis of more than 350 000 certified quality management systems within private and public sector organisations in at least 150 countries. ISO rules demand that its standards be reviewed periodically in the light of technological and market developments. The newly accepted revisions (December 2000) of the ISO 9000 series represent the most thorough overhaul of these standards since they were first published in 1987. The new versions take account of developments in the field of quality and the considerable experience that now exists in implementation. The new ISO 9000 is based on eight quality management principles: customer focus, leadership, involvement of people, a process approach to activities, a system approach to management, continual improvement, a factual approach to decision making, and mutually beneficial supplier relationships.

The emphasis of those involved in 'quality' has moved from 'detection and correction' to 'prevention' of failures. Today, the ISO 9000 standards are recognised throughout the world and are a key step in the removal of barriers to free trade.

Because the requirements of a quality management standard are generic and not as specific as product standards, different interpretations have been introduced for a number of sectors like the automobile and aerospace industries. The automotive industry e.g. developed a QS 9000 series of standards in 1995, which differ from ISO 9000 in detail, application and interpretations. The new -2000- revision of the ISO 9000 has the possibility to lay down such 'sector interpretations' as QS 9000 in "Industry Technical Agreements" (ITA's), which are developed by business parties.

One Management System

Since its introduction in 1987, the perception of what the ISO 9000 standards and guidelines are, has changed. The terminology used by those involved in the establishment, maintenance and auditing of ISO 9000 has changed from 'Quality system' to 'Quality Management system', to 'Management System' and – recently – back to 'Quality Management System'. This last move is to distinguish between management systems for Quality (ISO 9000), the Environment (ISO 14000) and health and safety (future ISO 18000). Interaction and overlapping is inevitable (Zwetsloot, 1994) and it is attractive for most organisations to develop one generic management system for all aspects (figure 5). Many companies are now asking for certification of their total management system including occupational health and safety, covering the current legislation and regulations. However one EN management standard, not to mention one international (ISO) standard, is not yet in sight to date. There is still no internationally recognised standard for occupational health systems, although there are initiatives ISO to draft guidelines in the near future (ISO/IEC 18000).

The establishment of a standard in general, and for an occupational health and safety system in particular, takes a lot of time. This is especially so because employers federations have been against the development of an occupational health and safety standard for fear of the proliferation of certification. However the new ISO 9000 (2000 version) is a formal management system, which defines the environment within a business and can also take into account requirements for an occupational health and safety system. It offers opportunities, for example through ITA's, to add health and safety requirements to the management system, even if there are no regulatory requirements or standards to date. This being the case, it is questionable whether there is a real need to have a separate standard for an occupational health and safety system, because it can be treated as part of (Total) quality.

The involvement of small and medium enterprises in the certification system is low which may be another shortcoming in the use of certification in health and safety. In The Netherlands a special certification approach, for small and medium enterprises, the so-called "growth certificate", has been accepted as a scope for accreditation by RvA. This provides a low threshold to help SMEs take the first steps in certification. This approach was developed in co-operation with the organisation of small and medium companies.

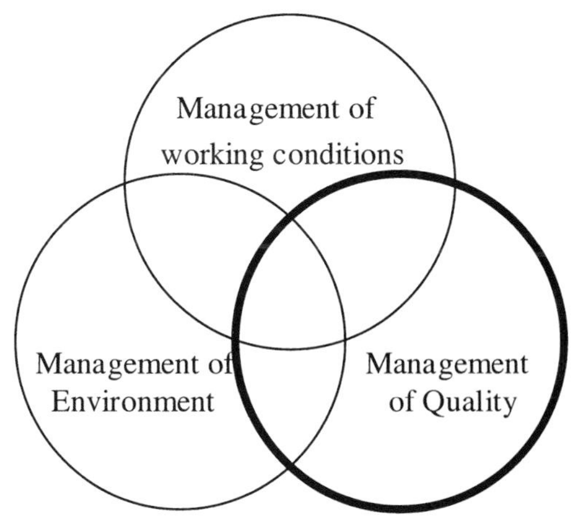

figure 5. Overlapping management systems

CONCLUSION

The question raised in this chapter was if third party certification is a tool for safety regulation. As stated, regulatory bodies are already increasingly using accreditation and certification as an impartial independent and transparent means of assessing that organisations are abiding by laws. Within the New Approach, certification to harmonised standards is accepted as a basis for presumption of conformity with the essential requirements. So the point of the question is not whether certification is used, but what the limits are in using it.

In general when certification is used within governance the objective could be defined as 'to provide social certainties with regard to the upholding of legislation' (de Graaf, 1996). The extent of the use of certification by authorities will depend among other things upon the ability of the parties concerned (the accreditation- and certification bodies and researchers) to justify, legitimate and explain the principles of certification. In particular, certification is and should remain a voluntary act; companies should be free to apply for it.

The main advantage for governments of using certification is its preventive character, the fact that all parties concerned are actively involved in it, and hence that it appeals to their own sense of responsibility. As certification of independent management systems moves towards certification of one integrated management system including quality, safety, health and environmental protection, this will contribute towards improved performance as well as reducing duplication and increasing effectiveness. Last but not least voluntary certification saves the taxpayer money, since it is the companies which have to pay for it.

The main disadvantage of using certification, for governments, is that, due to the decentralised interpretation of the law, the regulator may not be in possession of the best information to select priorities for health and safety protection. Where all parties concerned specify criteria, there is decentralisation of power because the interpretation of the law is 'shifting' to the certification process. Centralised government regulation makes room for a more decentralised interpretation system. This mental 'shift' is greater for countries with a Napoleonic system of law than for countries with the Anglo Saxon system. The Napoleonic system is highly centralised. The entire nation was linked together under national administration resulting in concentration of power, with law being interpreted centrally and then carried by appointed officials, civil servants and bureaucrats out into every village. The Anglo-Saxon law on the contrary has two separate and distinct branches, 'Public Law' and 'Private Law'. Public or civil law pertains to government-made law, which is strongly influenced in its interpretation by judges, through case-by-case jurisprudence. Private law pertains to the people acting in a law-making capacity, interpreted by local magistrates, resulting in even more decentralisation of power. Certification is in origin linked to the private law (of contract) and so is decentralised. However, it is now being drawn into the area of health and safety law, which is part of public law. The process of bridging these two areas is not always comfortable.

Government involvement in certification as one of the parties concerned in *centralised* 'Committees of experts', could lead to more effective and efficient laws to lower major accident risks. This is a more appropriate place for government to play a role than as the dominant party in the accreditation system. However, the principle that in these committees 'no single interest may predominate' may be difficult for some government authorities to accept. They may find it hard to accept changes coming from the 'market' without retaining some form of veto. This is due on the one hand to the legacy of the Napoleonic system and on the other to the uncertainty about the effect on public perception of greater influence coming from a wider range of stakeholders and about its political consequences.

Recommendations to the Use of Certification in Safety Regulation

- Enforcement and legal investigations can form no part of voluntary third party certification. They should be retained as separate powers and tasks of government.
- Codes of practise' should be established to guide the behaviour of certification bodies finding non-conformities with the legislation during audit or control visits.
- Under-representation of consumers and trade unions is a danger to the effectiveness of certification and undermines the applicability of certification in general, and in the safety area in particular. Action needs to be taken, largely in the form of financing, to enable such bodies to play their full part.
- The establishment of 'central committees of experts' is a method to make it possible for all parties concerned to participate in the certification process in a more cost-effective way.
- The integrity of the accreditation and certification structures must be assured. The mixing of conflicting business interests (certification/auditing and consulting) should be more effectively banned.
- The new 2000-version of ISO 9000 should be used to incorporate occupational health and safety issues into existing certification, without waiting for a separate ISO standard for that area.

References

Bennink, BJ.K. (1998). How to fit European Standards with the implementation of EU Directives in the context of Accreditation and Notification. *Contribution to EUROLAB Workshop* on Friday 26 June 1998. EUROLAB, Berlin.

European Commission, (1999). *Guide to the Implementation of Directives based on New Approach and Global Approach.* DGIII European Commission, Brussels.

Graaf de V. (1996); *Private Certification in a Governance Context*; Eburon Publishers, Delft.

Gundlach, H.C.W. (2000). Marking. *EOTC workshop 'No marking no selling?* Brussels, 13-14 Sept. 2000. EOTC, Brussels.

Gundlach, H.C.W. (1988). Certification bodies, their competence, evaluation and accreditation, *Paper contributed to the symposium 'On organising certification and testing for Europe'*, Brussels. DGIII, European Commission, Brussels.

ISO/IEC, (1992). *Certification and Related Activities. Assessment and verification of conformity to standards and technical specifications.* International Organisation for Standardisation (ISI), Geneva

ISO/IEC, (1994). *Guide 8402: Quality management and quality assurance; vocabulary,*

International Organisation for Standardisation (ISO), Geneva.

ISO/IEC, (2000) *9th Survey ISO 9000 and ISO 14000 standards*. International Organisation for Standardisation (ISO), Geneva.

ISO/IEC-9001 (2000), *Quality systems: model for quality assurance in design, development, production, installation and servicing*. International Organisation for Standardisation (ISO), Geneva.

Zwetsloot G. (1994), *Joint management of working conditions environment and quality*. Nederlands Instituut voor Arbeidsomstandigheden (NIA), Amsterdam.

15

INSIGHTS INTO SAFETY REGULATION

Barry Kirwan, Andrew Hale & Andrew Hopkins

INTRODUCTION

This final chapter attempts to draw some insights from the foregoing chapters, and make several fresh observations on the issue of safety regulation. The comments made here and the conclusions drawn are, however, entirely the responsibility of the authors of this chapter.

The chapter is in two main sections. It begins by revisiting the question whether regulation is really necessary and concludes that it is. It then asks the question why regulation of risk is so complex and tries to draw the lessons from the various chapters to illustrate and explain this, using the problem-solving framework proposed in chapter 1 as the structure for presenting a range of regulatory dilemmas or choices. The second main section discusses four residual, yet overarching issues, which try to present some new perspectives on regulation. These issues are first the need to have an evolutionary perspective for choosing regulatory regimes, matched to the particular evolutionary stage of the industry or activity. The second is the need to match regulation to specific situations based on a number of critical influences and a number of key factors affecting regulation are distilled from the foregoing chapters and from more general insights. This leads to a preliminary set of recommendations for achieving better safety regulation. The third issue concerns the lot of the regulators and their career prospects and the issue of staffing regulatory agencies. This is important because regulation will only be as effective as the competence of its staff and its resource base allow. Finally the need for better public risk communication is raised as an underdeveloped issue in the area of safety regulation.

In the first chapter we raised four questions which were debated in the workshop and in the chapters which form this book. It is useful to summarise here how far we have come in answering them.

The first question asked what the various forms of regulatory control were and how successful they had been and what dilemmas and advantages accrue from using them. The various chapters have provided a first description in safety science terms of the range of regulatory instruments met with in a range of industries and countries. The sample is certainly not complete and we have only descriptions, with little possibility to assess success or failure. What is certainly clear are a range of dilemmas and advantages experienced in using them. These will be summarised in the first part of this final chapter.

The second question asked how regulatory regimes are matched to the risk or activity being controlled and the third raised the related issue of what we can learn from one application and regime for use in other industries, countries or times. We shall propose in the second part of this final chapter an influence diagram which indicates some of the issues which appear to be important in matching the regulatory instruments and regimes to their targets. However this is very provisional at present and we have a great deal to learn yet about the success and failure factors in different regimes and applications. We have to conclude that, at present, changes to regulatory regimes seem just as likely to be driven by fashion and political issues as they do by any attempt to draw conclusions from assessing success or failure.

The final question asked whether regulatory regimes go through an evolutionary process. We have seen, particularly in the chapters by Hovden, and Hale et al, that evolution (or even revolution) takes place. We shall propose later in this chapter that that is essential for success, but we have to confess that we do not have hard evidence to prove this position.

1. Is Regulation Needed?

It appears generally from all the chapters that, despite the difficulties, regulation in some form is indeed necessary, or at least it is present. It is hard to think of any aspect of modern Western society in which the risks are not in some way regulated. Even in the sphere of sports and hobbies (from do-it-yourself to mountaineering) there are governing bodies or associations making safety rules and there are government or other safety regulations on the equipment which can be sold and used. There may be gaps in what is regulated, e.g. no competence testing before someone can take part, no safety management requirements, or no restrictions on when and where the activity can be carried out, but everywhere the regulator has at least a finger in the pie. At most we can ask where the regulator sits; is that in government or in a delegated third party or voluntary body? We may also ask the question whether society has become over-regulated and whether more should be left to the individual. Neither our book, nor the majority of research in this area, can answer that question clearly. Before and after studies of the effect on risk of safety regulation, or attempts at de-regulation, are almost unknown. Despite the talk of deregulation since the 1970's, the changes have been far more ones of relocating the place at which, and the people by whom, the rules are written, than of reducing the number of rules or the subjects they cover. Each accident or disaster that happens tends to result in media and political calls for tighter control from government, and seldom, if ever, for looser control.

Regulation is fundamentally intended to prevent harm coming to the public (including the workforce) and the environment. The papers in this book reinforce the need for some form of regulation. The assumption is that, without it, more accidents will happen, more frequently. Pure self-regulation does not seem to work, or at least does not seem to be able to have sustainable safety success. At the very least that is the perception which exists.

2. WHY IS SAFETY REGULATION DIFFICULT?

It is worth reiterating why safety regulation is an often inherently difficult area. First, there is the technology itself, which may be very complex and, even if not, its failure mechanisms and modes may not be well understood, documented, or even known. Second, such technology is always changing – sometimes there are large, fundamental changes, sometimes it is a matter of fine-tuning the technology to make it more efficient or of better quality. In both cases, however, new risks may be added, or else the likelihood of already known hazards may change and increase. Such changes in the 'risk profile' of an industry are difficult to anticipate, whether by the company or the regulator.

Third, there are cost aspects, such as market pressures, competition, the need for return on investment, etc, which all act as competing objectives for the players in the system, conflicting inevitably at times with safety. Such cost pressures vary according to the life cycle stage of the product, or system being developed. This life cycle aspect is a fourth reason that safety regulation is difficult. At the start of a new technological endeavour, existing regulatory systems or infrastructure may be insufficient or even wholly inappropriate to regulate the newly emerging and perhaps much-needed addition to the world's 'technological gene pool'. Later, as the industry is winding down, or under cost-cutting pressure, there may be reluctance to divert steadily diminishing resources to safety concerns that may be seen as overly cautious. There may be a feeling that investment in safety, when the factory, plant, machine or transport infrastructure is going to be phased out or replaced in a few years, is a waste of resources. "It will last our time", may be the cry.

A fifth reason safety regulation is difficult is the practical and logistical reason of having competent regulators available who are able to regulate. Regulation is a social function that needs people to achieve it. In an industrial age characterised and dominated by technological expertise, the human resources that comprise that expertise are often in short supply, whether for industry or regulatory bodies. In such a buyers' market it is vital to have answers to the questions whether the job of regulator is attractive in itself and whether it pays well enough. This leads on to the sixth main reason that safety regulation can be difficult, which is the nature of the regulatory relationship itself. The very nature of one person telling another what is and is not acceptable, is potentially confrontational, especially when the one has the necessary power to enforce his or her opinions. The nature of regulation itself can therefore contribute to a lack of trust, or an adversarial relationship between regulator and the regulated.

Regulation is difficult therefore, in terms of the technological nature of its object, the dynamic nature of open market systems, and the very human nature of the interactions that form the primary transactional medium between regulator and regulated. These difficulties will be discussed further in the remaining sections of this chapter.

2.1. What factors characterise regulatory performance?

In order to try to decide what types or forms of regulation best fit different types of industries, companies, and products, it is useful to try to elicit and structure the main factors influencing regulation efficacy, that have been mentioned in the preceding chapters. All of the following have been cited in some form or other and each word or phrase seems to capture an aspect of safety regulation, whether viewed from the regulatory, industrial or public perspective. We make a preliminary grouping here, but we freely admit that this is a very preliminary ethnographic sketch with many open ends. In the second part of the chapter we will attempt to place a clearer structure on the area by classifying the influences into a number of different categories.

A. Drivers and constraints

- Public concern/outrage
- Profitability
- Stability/maturity
- Rate of technology change
- Complexity
- Competition
- Out-sourcing
- Accident safety events
- Cost to taxpayer
- Size of company
- Safety competence

B. Socio-political framework or background

- Liberalisation
- Globalisation
- Existing regulatory cultures
- Legislative basis

C. Regulatory Traits/ Properties for effective regulation

- Power distance (power base)
- Independence
- Impartiality
- Integrity
- Transparency
- Expertise & resources

D. Styles of regulation

- Proactive vs. reactive
- Directive vs. Goal-based

E. Consequences for regulator – regulated relationship

- Trust vs. distrust
- Adversarial vs. cooperative relations
- Responsibility diffusion
- Complacency vs. vigilance vs. over-zealousness

The above list hints once again at the complexity and multi-dimensionality of safety regulation, and reminds us of its inherently sociological nature, even though it strains often to be seen as technical and scientifically-based. The categories (A-E) already indicate a certain ordering of influence, with the drivers and constraints interacting with the socio-political background to influence the choice of ways to combine the regulatory traits to produce styles of regulation which, in turn, have consequences for the regulatory relationship. In the next section we attempt to order some of these issues into a range of regulatory dilemmas, using as framework the problem solving structure set out in figure 2 in chapter 1, which is repeated here as figure 1.

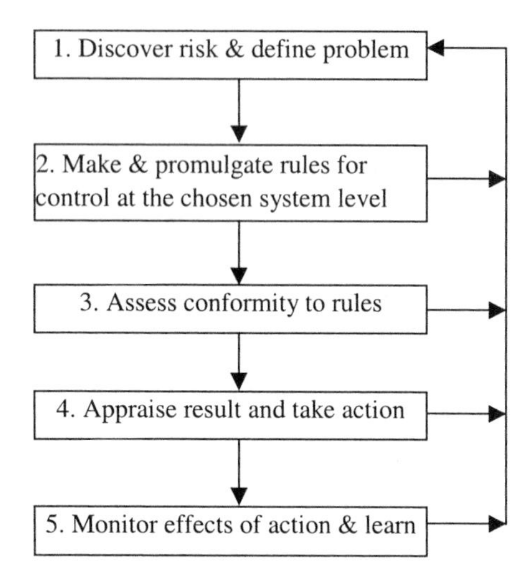

Figure 1. Regulation as problem solving process

2.2. Regulatory dilemmas

Although the authors in this book have not generally used this framework to order their chapters, we suggest that it can be a useful first approximation as structure for posing and researching questions related to safety regulation. The lack of an agreed framework for this subject area is an indictment in itself of the relatively undeveloped state of research and systematic discussion. We now discuss the choices for regulation under each of the five steps.

2.2.1. Step 1. Discovering and defining the risk.

The question in this step is whether this should be a task for the regulator or the regulated. Reactive regulation has always waited until there has been a manifest problem before imposing controls. The need for regulation has emerged from a disaster, a cluster of cases of occupational disease or serious accidents, or a public outcry. The pressure to become more proactive has led to the desire to anticipate such major problems. Much legislation requires the producer of the potential hazard to carry out research to test whether e.g. a chemical substance or machine is hazardous before it can be marketed. Safety case legislation also follows this line. Central to the European working conditions legislation is the requirement on companies to make an inventory and assessment of the risks of its activities as a basis for their risk control actions. The preference therefore seems to be to place this task on the creator and manager of the risk, not on the regulator.

However, some chapters (e.g. Williams, Baram) do raise the question as to whether this is satisfactory. Can we trust these actors to discover and reveal their own problems, when this causes them to incur the costs to control them? Should the regulator, especially in technologies that are new or uncertain, conduct and pay for research to discover and prioritise risks? Is this also a role for the regulator when the effects of an activity or technology are very diffuse, spread over a large population exposed at a low level?

2.2.2. Step 2. Making rules for control

We identified the options in chapter 1, to formulate rules at different system levels:

- technical-prescriptive rules operating at the direct interface between the hazardous activity and its controllers,
- goal-oriented rules providing those controllers with objectives within which to formulate their own rules,
- safety management rules imposed on the organisations exploiting the risky technology requiring them to formulate goals and/or prescriptive rules.

The choice of what is appropriate will depend on many factors and the search is to find which level, or which mix, provides the best way of defining and measuring risk control. What are the boundary markers we can most unambiguously define to indicate that the system is moving outside the envelope of control and needs corrective or punitive action?

Rasmussen (Rasmussen 2000, Rasmussen & Svedung 2000) argues that these markers will depend on the type of technology and its dynamics and hence on the dominant methods of keeping it successfully under control. He distinguishes:

- Technologies where failure is so catastrophic that the emphasis must be on regulating careful design to eliminate failure.
- Technologies where the emphasis can be on careful use and keeping the system within its design envelope
- Technologies where there is a possibility of intervening after the system has deviated and bringing it back to normal operation without harm.

He argues that the regulatory controls should match such distinctions. This is certainly true of the rules at the level of direct control of the risk, but it is not so clear if this is also true of rules for safety management systems and safety cases. It may be that the rules at this level are more generic. This was one of the questions posed in chapter 1 and at the start of the workshop. The fact that we have distilled some generic models here suggests that there is a set of principles for structuring regulation. However these are very general principles, comparable in abstraction to the principles of quality management to be found in ISO standards. We still have no clear answer to what the detailed emphasis and content should be for a given technology. Below are some considerations.

2.2.2.1. Prescriptive or safety management rules? The chapters confirm that there appear to be several main types of regulation. The first, and probably most well-known and applied, is **prescriptive** regulation, whereby the company is told what to do via compliance to a series of rules. This approach tends to engender a reactive response from the regulated companies or organisations. Prescriptive approaches are nevertheless widespread. They are perhaps most used for companies that do not have the resources, or do not choose to allocate the resources, to manage their own risks effectively in an anticipatory manner. This is particularly true of small and medium-sized enterprises.

The major alternative to the prescriptive or rule-based approach is the **goal-based approach**. This is where the regulator, for example, specifies targets for risk reduction to the company, who then have to reach those targets. Several chapters have shown goal-based regulation approaches – some of which have worked, and some which do not appear to have worked well enough (e.g. see Becker's chapter). This type of regulation is favoured in the UK nuclear power industry for example, and has been the dominant format for regulation in the UK since the Health and Safety at Work Act of 1974. This act led to more goal-based regulation, based on the idea that rules alone would not prevent accidents in complex evolving industries.

There has been a general move away from technical-prescriptive rules to goal-oriented and safety management rules. This forms the theme of the majority of the chapters. Critics of prescriptive regulation have usually argued on *a priori* grounds that it is inappropriate and, by implication, ineffective, particularly for rapidly changing or emerging technologies. These arguments may not be true of technologies and industries which are stable, well known and fully mature. Various studies reveal that, when enforced, prescriptive regulation is highly effective. The best such study was carried out in the US in the 1980s, using data on the enforcement activity of the federal OSHA inspectorate. Based on a sample of 6842 large manufacturing plants the research showed that plants that were inspected and penalised for violations of prescriptive regulations experienced a 22 per cent decline in injuries over the following three years. Inspections which did not result in penalties had no injury reducing effects (Gray and Scholz 1993; see also Boden 1985, Perry 1982).

Clearly, prescriptive regulation is potentially far more effective than is normally supposed, if enforced. In the US it was enforced. This suggests the intriguing conclusion that, insofar as prescriptive regulation has failed, it may have been for lack of enforcement, not because of its inherent unwieldiness. Lord Cullen recommended the abandonment of prescriptive regulation in the North Sea and its replacement by a safety case regime, in part because the inspectorate had failed to enforce its legislation (Cullen 1990). One is left wondering whether the Piper Alpha disaster would have occurred if the existing prescriptive legislation had been vigorously enforced.

One of the dangers of prescriptive legislation is that the regulator takes over the driving seat from the company instead of directing the traffic. If things then go wrong, the regulator is in the firing line of criticism. Such questions have been raised, for example, in the Netherlands in enquiries and in the media following two major disasters in the period of one year – the firework factory explosion in Enschede (Oosting 2001) and the New Year café fire in Volendam (Alders 2001). This led to strong criticism of the enforcement agencies and to a call for more enforcement of the prescriptive rules. The fear is also that the company will lean back and not take active thought or precautions to improve risk control over and above the regulated minimum.

In some systems we do not have the option to make rules about safety management. There is no clear management layer between the regulator and the direct controller of risk. This is the case with road traffic, at least for private motorists. It is also the case for self-employed persons, whether professional drivers or those hired in for work in the many industries which use sub-contracting, home workers and labour-only contractors. The only possibility for inserting a management responsibility in these cases is to make those hiring the labour responsible for the safety management. The chapters by Larsson, Hale et al, and Maidment all touch on this issue.

Goal-oriented regulation is sometimes described as self-regulation, since, while the goals and the general framework are set by government, it is up to enterprises to flesh out the detailed rules for themselves and to police their own performance with respect to these rules. While large enterprises are happy to carry out these functions, small businesses are not. They want to be told what to do so that they can do it and get on with the job, secure in the knowledge that they are in compliance with the law (Lamm, 1999). Ironically, then, many small businesses prefer prescriptive regulation.

As for safety management systems, various reviews of the literature conclude that their effectiveness has as yet only partially been scientifically demonstrated (Hale & Hovden, 1998; Gallgher, 1999). This is not to say that they are ineffective, but simply that there have as yet been too few studies capable of yielding convincing conclusions about the essential ingredients of a good management system in different industries and technologies. However, if the research on the effectiveness of OSHA regulation in the US is any guide, research on safety management systems will need to take account of whether or not their requirements are effectively enforced.

One of the main arguments for leaving prescriptive rule making to companies and not the regulator has been that government or other centralised rule systems, such as those run by social or private insurers, cannot keep up with change. Change is always a two-edged sword. It offers the opportunity of better control of existing hazards, but it also introduces new hazards and untried control methods. Several chapters discuss the response to this dilemma. Williams describes regulations in the nuclear industry which impose severe limits to change and restrictive procedures to assess its impact before it is allowed to take place. Baram describes the resistance to regulation in the rapidly developing industry of genetic manipulation, largely based on the fear that regulation will slow and damage this potentially lucrative technological change. One must, however, ask the question whether a technology would remain lucrative if all of the risks involved in its development and exploitation have to be internalised by the developers and exploiters, and, if this is not so, whether this is good or bad for society in the long run.

In summary it may be valuable to recall the proposals made by Hale & Swuste (1998), that we should consider the process of making rules as having three steps: formulating safety goals for the activity, specifying how to arrive at safety rules for it and actually formulating the specific rules. They argued that it was always necessary to go through these three steps or translations and that the only choice was at what level, from societal, through organisational to individual level, each translation should occur. If the people carrying out the tasks are responsible, highly competent, and self-aware, and if the activity does not affect others adversely, then all three steps can be done at the individual (or small group) level. In other cases, and particularly where harmonisation between individuals is vital for safety, it may be necessary to formulate detailed rules centrally (think of rules for which side of the road to drive on). Within this framework they also argue for leaving the detailed rulemaking as low as possible in the system.

2.2.2.2. What sort of laws? We have discussed the types of rules appropriate for safety regulation. However, there is another dimension that is ventilated in the chapters, namely what type of law safety law is or should be. In chapter 1 we took the position that it is a matter of administrative rules with the possibility of punishment (fine, stopping activity, etc.). However, there has always been a strong element of criminal law in safety law, notably in cases of gross negligence and wilfully exposing people to risk. Larsson, in his chapter, argues that safety law has been seen too much as industrial relations and employment law and not enough as public health law. This issue is touched on in other chapters. This debate is most important in considering how strictly the rules under the law are applied and tested, and what fora are used to decide upon disputed interpretations. Are these of the law courts, with their full panoply of the justice system, or are they much less formal tribunals of experts in one form or other? These issues return in steps 3 and 4 below.

2.2.2.3. Who writes the rules? A recurring theme in the chapters is the issue of who should be involved in writing or agreeing the rules which will be used as the basis for regulation. A number of variants were described in different areas of safety regulation, some related to product safety, and others to medical safety, or avoidance of industrial injury:

- Certification and accreditation
- Standards
- Safety groups/expert committees
- Individuals and companies
- Market-driven regulation ('free-market' regulation)[12]

[12] This is where the public stops buying the product or service that is unsafe. It is most evident in the area of food.

This list is intended only to show that, for many industrial areas, safety regulation is multi-faceted. Whilst an industry can have a dominant safety regulation 'mode' and actors, it is likely that there are a variety of safety regulation mechanisms working at different levels and locations in its various operations and inter-relations with suppliers, research laboratories, etc. Furthermore, these safety regulation mechanisms are not necessarily co-ordinated or orchestrated. Each exists where there is a niche. In effect, there is a safety regulation market-place, which itself is not necessarily regulated.

There has been a clear shift, in parallel with the move to self-regulation and safety management rules, towards involvement of the regulated in this process. In the certification system described in Gundlach's chapter this goes a step further, with the desire, but also the problem, of involving all interested parties, including employees and consumers.

An important question here is where the expertise lies to write (and to assess compliance with) the rules. Detailed risk control rules for a given technology can only be written by experts with deep experience of designing or operating that technology. Chapters such as Maidment's, Hale et al's and Williams' raise the question as to whether the government regulator has or can buy in such expertise. On the other hand de Mol questions whether the experts in the organisations running the technology or activity can be trusted to be independent enough to write their own rules. What mechanisms do we have for resolving this dilemma?

2.2.3. Step 3 Assess conformity to rules

An overriding concern in assessing conformity is whether the person or organisation which is required to conform is actually visible to the regulator. Larsson raises this issue strongly in his chapter on small businesses. These come and go so fast and are so shy of all forms of regulation, whether that is safety, environment, or tax, that they may escape notice completely. We might liken them to the shy scorpions, deadly, but hard to find. In contrast the major players in industry, the large companies and multinationals, are as conspicuous and easy to find as the major predators, the lions or the vultures.

Once the regulator has found the organisation and starts looking for the non-conformities, the question is how visible these are. Prescriptive rules at the level of direct risk control have the advantage that conformity should be relatively easy to find and assess. The problem is much more that there are so many prescriptive rules in a comprehensive set of regulations of this type, that the regulator does not have the time to assess them all. The choice of which ones to enforce can then seem arbitrary and unfair, depending upon the inspector's hobby horse, which punishes one firm for something that the company next door is allowed to get away with. There are also problems as to whether the prescriptive rules cover all eventualities and what should be done in cases where there is no rule, or where the existing rule is not appropriate or even dangerous for a given, perhaps crucial, situation. A safety net of general rules can resolve the former and interpretation by the regulator has to cope with the latter. With goal-oriented rules and ones at safety management level the first problem is much less, the rules are by nature more generic. However, the second problem is much greater. These rules are open to much more interpretation, which places a much higher demand for expertise and even-handedness on the assessor as well as a requirement of trust between regulator and regulated in which this interpretation can be discussed and resolved. This is particularly so for safety cases and safety management, where the regulated has to present and argue a case which the inspector has to judge on its merits.

An additional problem with procedural and management rules is the fact that companies can have well-defined and well-organised document systems which appear to comply, but which are not carried out in practice. Audits and compliance checks have to penetrate beyond the paper and test the implementation, which draws inspectorates back into the workplace inspection activities they may have thought they could greatly reduce when they moved away from prescriptive operational rules. The chapters by Oh, Hopkins, Hovden, Hale et al and Gundlach all provide illustrations of these dilemma, related both to government inspectorates and third party certification. The provisions in the new generation of ISO quality management standards, demanding continuous improvement in performance will merely exacerbate this problem. The various chapters discuss the possible compromises and combinations between prescriptive and generic rules which can resolve these dilemma. Maidment, for example, describes the cascade of safety cases at different levels in the railway system in the UK, which attempts to regulate the translation from generic management systems to specific rules.

A final issue under this heading is the access to privileged information to assess conformity. A number of chapters address this issue in relation to the trust required between regulator and regulated and the potential conflict between the regulator's duty to inform the public and the requirement to see into the innermost, intimate secrets of the organisation in order to assess compliance.

2.2.4. Step 4 Appraise the result and take action

Appraisal implies priority setting and the assessment of what action would produce the ultimately desired end result. Here the philosophies of criminal law and other forms of law clash most clearly. Under criminal law violation leads to punishment as end result, with comparatively little concern about whether it also leads to improved behaviour in future. Within a public health perspective the paramount aim is improving the total of public health and well being and punishment is seen as means and not end. This clash is most clear in deciding on actions after an accident which has caused injury or death, particularly where there are multiple fatalities. There is often intense public concern and a strong inclination both to blame and to demand retribution. A prosecution is likely in these circumstances, the target normally being a corporate entity. The prosecution is likely to be for a regulatory violation, not for the harm caused by the violation. This is obviously so where prescriptive regulation prevails but it is also the case where goal-directed regulation is in force.

Even in these circumstances, employers are strictly speaking prosecuted for their failure to ensure that workplaces are as safe as practicable, not for the death or injury which may have resulted from this failure. However, it is striking that prosecution in most countries and industries is much more likely when major harm has resulted, than in a near miss incident or a 'simple non-compliance'. In the public mind in particular the prosecution will be for the harm done. Thus the prosecution for the regulatory violation is being called upon to serve a function which it is not designed to serve – to give expression to the community's sense of outrage and to provide retribution for causing death or serious injury.

This can generate feelings of unfairness all round. Because regulatory violations may not be regarded as serious offences in and of themselves, that is, in the absence of harm, the penalties imposed may not be commensurate with the harm, and the public may feel that corporations have been let off lightly. On the other hand, major incidents are always accidental, in the sense of being unintended and unwanted events, and to blame the corporation often seems unfair, at least from the corporate point of view. The problem stems ultimately from the point noted at the outset of this chapter, namely that it is not possible for safety regulation to prohibit harm. One possible solution to this problem is to make use of the traditional criminal law in these circumstances, rather than the apparatus of regulation. The traditional criminal law makes it an offence to cause harm (death or injury) intentionally or because of serious negligence. The use of the traditional law avoids both the problems outlined above. On one hand, a prosecution cannot be initiated simply because some regulatory violation resulted in harm; it needs to be shown that serious negligence was involved. This acts as a brake on "unfair" prosecutions. On the other hand, where a corporation has been seriously negligent, penalties can be much higher than those provided for under most regulatory law. The problem is that the criminal law is traditionally aimed at individuals, while the defendants envisaged above are corporations. Attempts to prosecute corporate entities for criminal offences have proved difficult, highlighting the need for law reform in this area (Wells, 1993). The failure of the prosecution of P&O for manslaughter following the Zeebrugge disaster, for example, has prompted the British government to reformulate the criminal law to place the offence corporate manslaughter on a firmer footing.

Retribution is not the only purpose of prosecution. Deterrence is another. Insofar as prosecution can be shown to *deter* negligent behaviour, it can be justified on public policy grounds. Whether prosecution following an accident in fact functions to make companies attend more conscientiously to risk is an empirical question. And as is so often the case, there has not been sufficient research to give a conclusive answer to this question. It should be noted, too, that court sentences can include detailed requirements that the offending corporations repair the defective management and information systems which facilitated the offence (Fisse and Braithwaite 1993). This is an under-utilised judicial power, but one which serves the purposes of accident prevention well. It resembles closely the market-driven powers, which insurance companies already have, to demand detailed improvements as a condition for giving or renewing insurance cover. The power of the courts, however, exceeds the power of the insurers in this respect, especially where competition between insurers allows the insured to pick and choose whose rules to accept.

In summary, while the possibility of prosecution following accidents may sometimes impede organisational learning, it may also encourage more conscientious risk management. It is clear that, even from a dispassionate accident prevention point of view, the task of designing sound policy for prosecution following accidents is not straightforward.

In a sense, the damage claim system, particularly in jurisdictions such as the USA where punitive damages are commonplace, fulfils this requirement of retribution against the perpetrating organisation. However, the costs of such a system are high and victims may have to wait many years before compensation is paid to soften their sufferings (Larsson & Clayton 1994). This public need to express outrage and take retribution is quite clear in the aftermath of major disasters. It is especially likely where there is a perception that the regulator has failed to act effectively, and where the issues at stake have strong moral and ethical overtones, as Baram discusses in his chapter on gene manipulation.

The preceding discussion has focussed on the prosecution of corporations. In some jurisdictions regulations hold senior managers and directors responsible for regulatory violations, if they have not exercised sufficient care to ensure that safe procedures are being followed. These provisions are seldom used, except where the individuals concerned have been particularly negligent with respect to their obligations, but they do appear to have a dramatic publicity effect.

2.2.4.1. To punish or persuade? Let us now turn to the case where non-compliances are detected in the absence of direct harm. Inspectorates face an enduring dilemma when confronted with such regulatory violations: is compliance best achieved by prosecuting and punishing these violations, or is the best policy to persuade violators, by means of exhortation and warnings to bring their practices into compliance with the law? (Braithwaite, 1985). This question has dominated much of the regulatory debate. Research has demonstrated that there is a systematic variation in the position of inspectorates along the punish/persuade continuum. The general pattern is that "the greater the relational distance between the regulator and the regulated and the less powerful the regulated, the greater the tendency to use formal sanctions" (Grabosky and Braithwaite, 1986:217). To illustrate this idea: the police are at a considerable relational distance from most offenders they deal with, and they are considerably more powerful. They thus tend to react punitively to non-compliance. On the other hand, where inspectors have close relations with a small number of regulated companies, and particularly where these companies are large, they tend to be less punitive. Many specialised safety inspectorates fall into this category as the chapters on the nuclear, offshore and railway regulators demonstrate.

Quite apart from the issue of where inspectorates in fact stand on this continuum is the question of where they *should* stand. The debate in relation to safety inspectorates has been intense. Where an inspector identifies a violation of some prescriptive regulation, or alternatively of some code of practice or guideline, and where no harm has occurred, as is typically the case, there is a strong tendency to regard this as a "technical violation only". Many safety inspectorates prefer to give offenders numerous chances to comply in these circumstances (Hutter 1997). The critics are scathing of this approach, accusing inspectorates of going easy on white-collar criminals (e.g. Carson 1970). On the other hand, occupational health and safety inspectors in the US have had a tradition of penalising such "technical" violations and have been accused by *their* critics of "regulatory unreasonableness"(Bardach and Kagan,.1982). It is clear that inspectorates are in a no-win situation, which again results from the nature of the regulations they are required to enforce. (For a more detailed discussion of this issue see Hopkins 1994.)

Academic commentators generally argue that the solution to this policy dilemma is an "enforcement pyramid" (Fisse & Braithwaite, 1993). The suggestion is that regulators should begin by using the tactics of persuasion. Most matters can be dealt with in this way, at the bottom of the pyramid, as it were. However, where resistance is encountered, or deliberate and/or serious offences are uncovered, regulators should be willing to escalate rapidly up the pyramid and in extreme cases impose very heavy punishments. Few cases would need to be dealt with at the highest level – hence the pyramid shape. This discussion takes no systematic account of the punishment imposed by civil proceedings related to liability, which may be many times, at least in financial terms, that at the discretion of the regulator and/or the prosecutor in criminal proceedings. The reconciliation of these two sources of punishment remains an open issue.

2.2.4.2. To advise Both punishment and persuasion involve the exercise of authority. But this does not exhaust the role of the regulator. In many contexts, inspectors have another function, to provide expert advice, or to educate (Kagan and Scholz, 1984). Traditionally inspectorates have seen a part of their role as collecting and disseminating best practice in order to apply a constant pressure for improvement across technologies (Hale 1978). Larsson argues in his chapter that this advisory role has gained prominence in modern inspectorates. Moreover, where inspectorates have been expected to generate revenue, as in some Australian states, they have charged for their advice. This has led at times to focus on customer satisfaction, to the detriment of the enforcement function. In contrast, the certification system (see the chapter of Gundlach) regards advisory, compliance testing and enforcement roles as strictly incompatible and goes to quite considerable lengths to separate them. It assigns enforcement to a government regulator, as a function impossible for a private organisation to do. It prohibits the same organisation having a consulting and advisory arm and a certification arm and imposes limits on auditors certifying companies they have advised in the last few years. This independence between advice and punishment may be desirable from such a point of view, but does create problems in practice. For example it hard to know where to recruit and how to retain enforcement staff who have a sufficient depth of knowledge of an industry to regulate it in a sufficiently flexible way to cope with goal-based regulation or exceptions to prescriptive rules, if they cannot have advisory roles too. Not everyone finds the role of high-level 'policeman' attractive. We return to this point later.

Self-regulation implies a different role for inspectorates, not necessarily a reduced role. Unfortunately the move to self-regulation has often been accompanied by a reduction in resources available to inspectorates. Of necessity this has meant that small businesses, which rely on inspectorates to tell them what to do, are being left more than ever to fend for themselves. In the small business sector, therefore, self-regulation in practice has often amounted to de-regulation, that is, the withdrawal by the state from the control of risk (Berger, 1999). As Larsson points out in his chapter, this is particularly disturbing, given that outsourcing and subcontracting means that workers in modern industrial economies are increasingly employed in small enterprises and that most workplace injury occurs in this sector of the economy. One way in which regulatory authorities have responded to this challenge is by encouraging various sorts of networking among small business employers. This is seen as a way of disseminating information about OHS and encouraging them to engage in rudimentary risk management activities (Mayhew, 1999). However surveys routinely show that small employers regard accidents as being beyond their control and that they hold their workers responsible for their own health and safety. As long as these attitudes remain unchallenged, there seems little hope that anything can be done about what Larsson calls the pulverisation of risk and the privatisation of trauma in the small business sector. Other regulatory jurisdictions (e.g. the Netherlands) have made it a requirement that small businesses sign a contract with a private working conditions service for expertise on risk assessment and prevention. In this way the advisory and enforcement tasks of regulation have been split.

One of the incentives which has been applied to encourage companies to self-regulate (and to escape from strictly enforced prescriptive rules) is the promise by the inspectorate that they will then be left much more alone and not inspected so often. This policy raises two interesting questions for research (which has also not yet been done). Do companies left alone in this way indeed make their own rules and apply them? Is it rational to withdraw inspection just on the proof that a company has the structure to regulate itself and the competence to do so – or should the test be of the commitment to use these for self-regulation? If the latter, how do we measure commitment without conducting the sort of detailed inspection which the companies are being promised they no longer need to undergo?

2.2.5. Step 5. Monitoring of actions

This step does not receive much attention in the chapters. An ongoing relationship between regulator and regulated is, however, an essential feature of a system which wants to learn and improve. As Gundlach describes, the certification system has this monitoring role built into it with formal checks at intervals between full certification visits.

Inspectorates may also formalise this follow-up. In a number of cases a monitoring element is added to the enforcement step. After first discovery of the violation or non-conformity the organisation is given a fixed time to put it right. If it has not done so then it faces rapid punishment, formally for not complying with the order to improve, rather than for the original offence. This represents a way of shifting up the pyramid of enforcement mentioned earlier (Fisse & Braithwaite, op. cit.). This breathing space between discovery of non-conformity and publicity about it is also a factor in the discussion between regulators and companies about access to the companies own audit and improvement records. There is much less resistance to making these available to inspectors if the company is given, or has had, the time to put the significant shortcomings right. It is then not hoist on its own petard of being punished for problems it has revealed itself.

3. Residual Overarching Issues

3.1. How do regulatory systems evolve?

One aspect that is not fully addressed in any chapter, but is perhaps most evident in Hovden's, is the need to adapt safety regulation to the stage of the industry or company in the life cycle. However, this is not simply the traditional 'system design life cycle' that is being discussed, but rather it is the '*system in the market*' life cycle. Thus, for example, a new endeavour or technology may initially be borrowing high risk capital to get itself 'off the ground'. If successful, it will then need to give a good return on investment, but also simultaneously to re-invest in order to consolidate its position in the market-place. It may then go through a steady or even increasing period of growth until it matures. Later on, it may diversify, and ultimately it may decline and need to shut down operations. Alternatively, the endeavour may fall foul of situations outside its control, such as collapsing markets or sudden competition etc., leading to a dramatic fall in profitability and in turn to premature decline or cessation of trading and/or operational activities.

This example suggests that there are many periods during the life cycle of the system-in-the-market when safety is unlikely to be the major priority. Unfortunately, these periods may coincide with times when safety is most threatened. For example, at the outset of a new system, technology, or company, there may be most uncertainty and lack of knowledge about risks. This is a period when 'latent errors' or 'resident pathogens' (Reason, 1990) can take root in a system and begin their incubation (Turner, 1978) towards a later accidental outcome. It may also be a time when the industry is dominated by 'cowboys' in for a quick profit. Much later, the system may be getting old, complacency about risks may have set in, and resources will be diminishing. There will be a demand to extend the life of plants and structures beyond their original life span and design envelope, in order to squeeze out extra profit. In between there may be a period of relative stability during which safety can receive more appropriate investment and safety regulation can become mature. During such a period earlier mistakes or oversights can be rectified, and plans can be laid down for future less prosperous times.

This suggests that some way of enhancing safety knowledge and practices for all phases in the life cycle of such systems, and thereby developing appropriate control mechanisms, needs to be developed. Baram's paper shows how, in completely new areas, this need can fail to be satisfied.

A key question for the company and regulator alike, is what happens when there are outside forces that seriously threaten not just profitability, but the very survival of a company. It is difficult in such an environment for a regulator to press too hard. If there are clear safety issues and imminent dangers, then the regulator's responsibility is also clear. But when such dangers are not evident and pressing, the hazards are 'distant' and their likelihood stretches credibility, there may be an implicit relaxing of normal safety goals from the principal side of the regulatory 'fence', particularly if the regulator-regulated relationship is 'cosy'.

It is difficult to derive solutions for these life cycle considerations. However, three potential insights are as follows. First, prescriptive regulation may not be the best form of regulation to begin with, even though it is the most obvious one. Or, rather, prescriptive safety regulation on its own may not be the best format. This is because it allows the company to delegate (and possibly relegate) safety concerns to another party while it focuses on getting the company going. This does not engender safety culture in the formative stages of a company or technical endeavour. This allows problems to set in early on, and can reinforce a return later to a '*safety-yes-but-not-now*' culture, if the going gets tough for the company. We need to look for ways early in the life of a technology for forging industry-regulator partnerships to tackle these problems, rather than allowing them to become the subject of conflict. If such partnerships are not possible, or the developers of the technology do not respond to the demands for control, we are left with many questions how best to ensure that the regulator intervenes to stop abuse without smothering the new industry in its cradle.

The second insight is that there may be a need for *rapid regulatory prototyping* for new technologies etc. This means some way of rapidly scoping the safety potential of new endeavours or processes or technologies, in a far shorter time frame than conventional legislation will take. The third insight is that for most life cycle processes, there will be a period of relative stability when safety can really be consolidated in a company, and this period must not be squandered. This is when the company and regulator can most usefully collaborate to achieve a common goal – sustainable safe production.

These are relatively high-level insights, but may be food for thought to some regulators. In any case, what is clear, is that the regulator needs to be aware of, but not overly constrained by, the wider market picture and how it is affecting the system being regulated.

3.2. Influencing factors revisited.

In section 2 of this chapter we listed a very large number of factors relevant to safety regulation and to choices within it. Figure 3 attempts to structure some of these aspects and the insights from the rest of this chapter into an influence diagram.

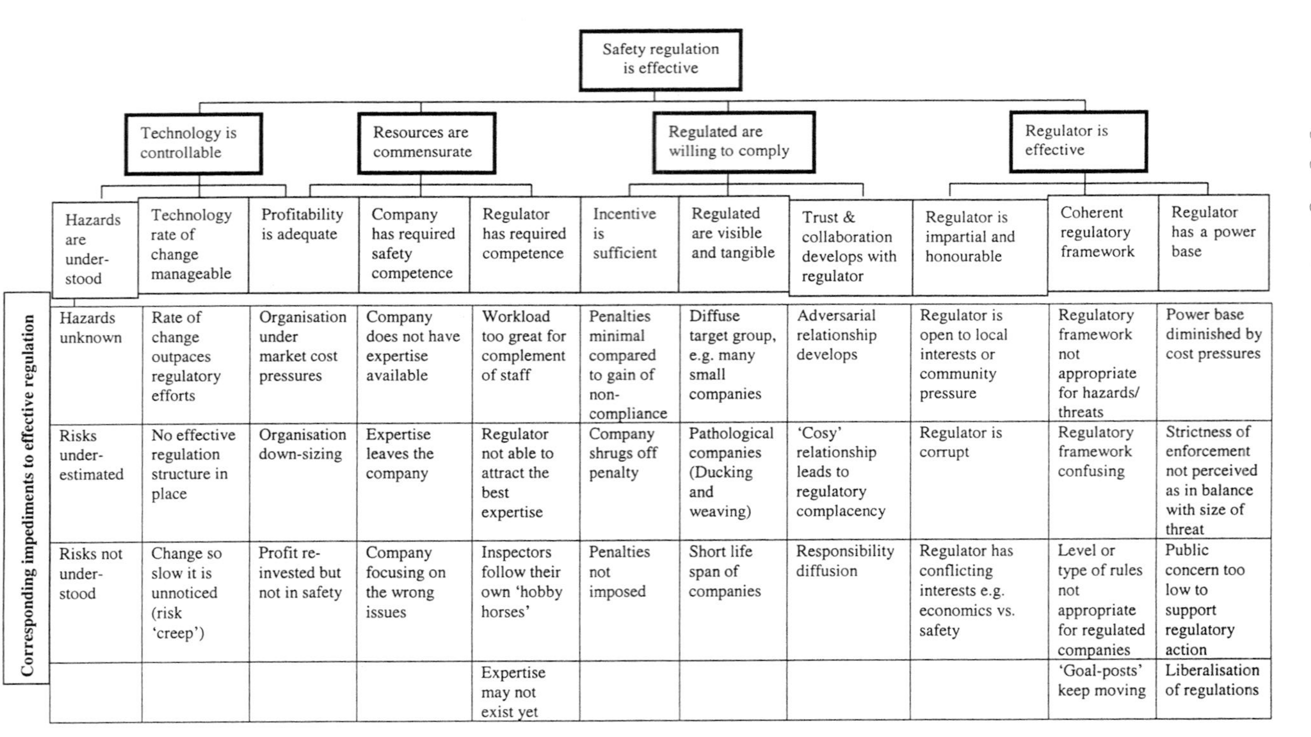

Figure 3. influences on regulatory effectiveness

An influence diagram is not intended to model all factors influencing a complex system, only those that are significant (it is called a 'requisite model' in the field of decision analysis: see Humphreys, 1988).[13] This is a top-down structure, with safety regulation efficacy at the top level. According to this diagram, which would ideally be based not on 'arm-chair analysis' of a group of chapters, but on data collected from the real world or via expert elicitation methods, there are four main factors influencing safety regulation success or failure. These intermediate factors themselves are then decomposed further into more specific influences. Beneath this level, problem areas are cited for each third level factor that can lead to unsuccessful safety regulation.

3.3. Towards better regulation: some provisional recommendations

Finally, a number of bullet points per heading try to summarise the insights which the authors of this chapter have gained from the whole book and the efforts to draw it together. They are organised under the main headings of the influence diagram

Technology Controllability

- *In dubio, abstine* – if there is significant uncertainty about the nature or effects of a process, and if appropriate tests or facilities or control mechanisms are not available, then it should not be accepted. [Absence of evidence is not evidence of absence: the onus is to prove something is safe, not to accept it as safe {hope it is safe} until events prove otherwise.]. It is recognised that this 'precautionary principle' is a very tough criterion. But the regulator's primary duty is to the public and the environment, and this principle logically flows from this prime responsibility. In practice, if this principle is adhered to by the regulator as a 'default' assumption, the required tests or facilities will soon emerge. Indeed, it may encourage more thinking ahead in safety terms, enhancing safety culture. This hard line needs to be adhered to at least until the following point has been satisfied, but it can only be kept to if there is international agreement on it as principle and hence it imposes no selective competitive disadvantage.
- There needs to be a form of 'regulatory prototyping', a regulatory research body able to deal with emerging technologies and rapidly assess their regulatory needs. Such a body must be independently funded, avoiding conflicts of interest.
- There needs to be a way of fast-tracking regulatory legislation for new technologies or applications.

[13] An influence diagram can in theory be quantified to give an indication of the likelihood of the 'top event' being true, e.g. the likelihood of safety regulation being effective for a particular industry. However, there is no intention here to quantify, merely to use the influence diagram format to highlight what is needed for effective safety regulation, and what can detract from it.

- Regulatory bodies for new industries, technologies and applications, representing new organisational 'cultures', will require a significant influx of new regulators (i.e. not from other regulatory organisations and existing cultures).

Resources

- There should be more external and generic training on how to be a regulator – it should become recognised as a career path. It should lead to a recognised qualification, preferably certified at least at the same level as the health and safety professionals who are required by law in many European countries. In addition to the generic training in regulation and prevention the regulator needs sufficient knowledge of the technology and industry being regulated.
- The regulator needs the time and resources to regulate, to 'get beneath the surface'.
- Regulator activities need to include on-site inspection in order to penetrate behind fine words and handsome paperwork..
- The regulator needs appropriate expertise – e.g. a specialised area of expertise, but also a broad understanding of other areas, to prevent 'hobby horses' being ridden, and to see the potential unsafe interactions that can occur in an industrial concern. This implies that the task of a regulator is one at graduate level.

Regulatory Cooperation & Willingness to Comply

- The regulated need to be, or be made, visible. This is the major challenge of the SME and of the trends towards downsizing and outsourcing. It is also a challenge in fast-developing technologies or industries going through periods of severe competition and 'shake-out'.
- The regulated need to be responsive to regulation. The processes of privatisation have raised important questions on this issue which are developed further at the end of this section.
- Penalties and the probability of being caught, particularly for pathological companies need to be meaningful and commensurate with the gain from non-compliance
- Positive incentives for compliance need to be built in where possible
- The regulator needs to be seen to be fair (and 'firm'), with no favouritism between firms, or 'hobby horses' which deviate from accepted practice.
- There needs to be a degree of 'distance' between the regulator and regulated, at least at some levels in the respective organisations, to avoid 'cosy relationships'.
- There should be frequent consultation between regulators and the regulated or their representatives. This poses particular difficulties for poorly organised or diffuse industries
- Regulatory processes should be relatively transparent.
- Safety boards are useful formats for enhancing collaboration – such groups need to comprise a broad and balanced cross-section of stakeholders, including trade unions, consumers, etc.
- Evidence of pathological relationships between regulator and regulated should signal the need for consultation and analysis of the relationship and possible re-direction. Such pathologies should not be allowed to persist and become the status quo.

- Quis custodet ipsos custodes?[14]. There is a need for rotation of regulatory tasks within an inspectorate to avoid undue influence and a cosy relationship developing. Independent review of the work of regulators by an ombudsman or by peer review between ministries or even across countries[15] should be established.

Regulator Effectiveness

- Regulatory philosophy and practice need to be clear and coherent.
- The regulator's remit needs to be clearly and publicly explicated
- There needs to be a degree of regulation 'stability' and predictability within a particular industrial sector.
- It may be advisable to separate 'prevention' (rule-making) and 'enforcement' functions in the regulatory body, but there needs to be particular attention to closing the learning loop between these two, so that rules are both feasible and enforceable.
- There needs to be regulatory coherence at an international and indeed global level. Different standards and requirements in different geographical locations clearly sends a wrong signal to industries who operate multi-nationally. Such international standards must not however be 'diluted' and reduce requirements and standards to the minimum that the least developed country or safety framework can manage. There needs to be scope for lead countries and followers in a progressive approach to regulation world-wide
- The regulator needs a power base. This consists firstly of the power to shut down an industrial process, but can also draw on the power to prosecute, to impose a high fine (particularly for an SME) and to threaten the image of the company through adverse publicity (particularly for a large multi-national). Competition between certifiers of management systems can threaten this power base, since one may be disinclined to impose hard conditions for a certificate for fear that the customer will go to a rival more prepared to be flexible.
- The power base needs cognisance of, but independence from, the profitability of the industry being regulated – i.e. the regulator must be aware of cost pressures and their impacts on the industry, but only to an extent, and not to the detriment of safety.
- The regulator must not have conflicting interests.

The above are not comprehensive, but it is believed that they are coherent (i.e. they do not contradict each other). They can provide at least a basis for choosing what sort of regulation is appropriate for what type of industry and for searching for appropriate ways of removing or coping with the weaknesses found.

[14] Who regulates the regulator?
[15] In aviation this practice is already established on a voluntary basis to assist developing countries. It is also standard practice between accreditation bodies (see also Gundlach's chapter).

Most of these issues are self-explanatory, or have been dealt with sufficiently earlier. We say a few more words about the issue of privatisation at this point, since it comes up many times in the chapters. The privatisation of government-owned enterprises is a prominent feature of government policy in many countries. In the 1980s it was common to speak of privatisation and deregulation as going together, but it is now clear that privatisation has necessitated considerable regulatory growth. New regulatory agencies have been necessary to ensure that the public interest is protected whenever it is threatened by the commercial imperatives driving the new organisations. Even so, the evidence suggests that the transfer to the private sector of many services - public transport, water and sewage, telephone, and so on - has seen significant declines in the quality and reliability of these services (Neutze, 1997, Cullen 2001)). Where privatised enterprises have the capacity to cause major accidents, the question of the appropriate regulatory response becomes critical.

The matter is particularly salient in the light of recent rail disasters in Germany and in the UK (Cullen 2001), both in the wake of the privatisation. Maidment argues in his chapter that the regime regulating rail safety in the UK was actually strengthened at the time of privatisation and that neither the Southall nor the Ladbroke Grove accidents can be attributed to any decline in safety standards following privatisation. On the other hand, Becker argues that safety in German rail system did decline in the wake of privatisation and that this was one reason for the high-speed train disaster in which 102 people were killed at Eschede in 1998. Although these authors differ in their explanations for the accidents they consider, their analyses confirm that the privatisation of hazardous enterprises calls for a special regulatory response to counteract newly introduced commercial pressures.

3.4. The regulator's lot

Norros and Reiman's chapter usefully elucidates the complex nature of being a regulator, which can be seen as requiring a difficult balancing act, and one with little gratitude to be expected. Regulators also appear to have a somewhat abbreviated career horizon.

Regulatory authorities appear to be atypical (and sociologically interesting) organisations, in that they comprise a very high proportion of experts with little obvious ‘apprenticeship culture’. This raises a number of serious questions: how do ‘experts’ become regulators? What do they gain and what do they lose in this process? How do they maintain their expertise, especially if research or application-based? How do non-experts become regulators, and manage to do their job effectively when the regulated may be continually ahead of them? None of the chapters has dealt in depth with these issues, though Hale et al do raise it when they discuss the problems of reorienting inspectors from a technical to a management-based inspection approach.

The options for recruiting regulators are limited. Some recruit graduates (or lower level staff) directly from their first qualifications and give them on-the-job training to familiarise them with industry. This usually results in criticism from the regulated that the inspectors have too legalistic a view of regulation and are too inflexible in the face of the practicalities of compliance. The alternative is to recruit from those with industrial experience, either as managers, technical experts or health and safety advisers. There may even be a more frequent shifting between inspectorates and the industries they regulate, with career moves or temporary detachments. Different countries also favour different training regimes, some combining courses for regulators and experts from the regulated, some keeping them strictly separate. We know of no studies which examine the pros and cons of these different training and career paths. The basic knowledge of hazards, risk analysis and prevention needed to be a regulator or to be an adviser of the regulated is clearly the same, but the skills of developing and applying regulation may differ somewhat from those needed by the adviser to the regulated in producing organisational and technical change.

Regulation can and perhaps should be seen as a craft. It certainly appears to require some exacting aptitudes, particularly when interacting with the regulated. There could be more consideration of regulation as a career path, and appropriate selection (regulators seem to have a particular value structure – see Reiman and Norros), training and accreditation to accompany this 'craft'.

A recurrent theme in this chapter and in the whole book is the issue of the relationship between the regulator and the regulated. The very existence of a regulator implies a degree of mistrust and antagonism. De Mol talks of the paradox that without this mistrust regulation would not be necessary, yet successful regulation relies on a sufficient degree of trust in that relationship. Trust determines the willingness of organisations to allow access to their information and records. Trust underpins the process of interpreting generic rules for particular circumstances. Trust provides the basis for acceptance of the regulator's proposed action to punish or recommendations to improve the risk control. Do the regulator and the regulated form part of one system with a common aim of furthering the technology and industry within agreed safe boundaries? Or is the relationship an adversarial one, with one side fighting for safety and the other pursuing profit, prestige and survival? Should the inspectorate temper its demands for safety if they would threaten the economic viability of the industry or drive it abroad? The balance on these dimensions differs in the different chapters. The ones relating to the nuclear industry (Williams, Walther) tend to show more commonality of interest. Those relating to the offshore industry (Hovden, Hale et al) show a sharper conflict with serious questions raised about an inspectorate dealing with safety being part of the same organisation which promotes the development of the industry. The small businesses (Larsson), the new technologies (Baram) and the newly regulated sectors (de Mol) show the greatest antagonism between the two parties.

Related to the issue of trust is that of independence. Are the rule makers and regulators sufficiently able to adopt a critical view of what they regulate, or are they too close to, and cosily in cahoots with the industry? Baram and de Mol indicate that this lack of independence makes professionals in an activity unsuitable as rule makers and enforcers, and self-regulation problematic. Maidment argues that any small, specialised and tight-knit industry has difficulty in finding independent people with sufficient knowledge and understanding of its workings to be able to assess its safety. He advocates the use of international experts as one way out. Traditionally another solution has been to buy in mid career experts or to use consultants. Both Williams and Oh describe this approach and its shortcomings. Gundlach accepts that this is also a problem with the certification system, even though it was set up to guarantee independence and goes to great lengths to do so. Because of competition in the certification market a certifier may be too reluctant to interpret rules strictly or to withdraw a certificate for non-conformity.

The growing importance of this question of independence is also seen in the trend to establish independent accident investigation boards, notably in transport safety. However, the recent disasters in the Netherlands have prompted the decision to extend the remit of such a board to all types in incident and disaster, adding industrial safety, agricultural disasters such as foot and mouth, food-related concerns such as genetically manipulated crops and even civil disturbance and public safety incidents.

As Hovden points out, we need to analyse this relationship in terms of power. What power does the regulator have over the regulated? He argues that part of the success of the Norwegian Petroleum Directorate was due to their power to grant or withhold licenses for exploiting blocks of the North Sea. Independence which leads to powerlessness is then a bad thing.

The chapters by Oh, Williams, Reiman & Norros and Hale et al. in particular concern themselves with the implications for the staff of an inspectorate of these choices among alternative regulatory regimes. The issues which are raised centre around the need to ensure a combination of different skills. There is the deep expertise to make rules and assess conformity with them, which must be coupled with the broad perspective to see the wood for the trees, establish priorities and spot things which the regulated are too close to be able to see. Acquiring and keeping the competence and practical experience of the regulated activity is a considerable problem. We noted above the possible solution of buying in expertise from the regulated industry, but this is limited by the buying power of the regulator, who may not be able to offer the salaries and challenges necessary to attract the best recruits.

The move from prescriptive regulation to the safety management system requirement has seen a profound change in what is required of regulators. Under prescriptive regimes regulators functioned largely as inspectors, operating mainly in the field and checking whether employers were in compliance with detailed legislative requirements. However, where the focus is on safety management systems, regulators function more as auditors. This involves spending much more time in the office, examining paperwork and, in the case of hazardous facilities, reviewing risk calculations. Time in the field is then focussed on verifying that the system is operating as intended. This auditing activity involves a very different set of skills and many inspectorates have had difficulty retooling and retraining to deal with the new way of working (see e.g. Hale et al's chapter).

Regulators also worry that because they spend less time in the field they tend to lose their intimate knowledge of the facilities they are responsible for and therefore are less able than previously to identify the signs of deteriorating safety standards. Expertise gets out of date if not practised for a few years and inspectors lose the "gestalt" or "situation awareness" to be able to recognise ill-defined problems and get to the heart of the matter. Becker's chapter makes some interesting suggestions about how this might be dealt with.

The various roles described in step 4 above create tensions when they have to be carried out by the same person. Reiman & Norros consider these tensions in detail, but they are also apparent in the chapter by Larsson. We have explored in the earlier sections some of the possibilities of splitting these roles among different regulatory agencies in order to reduce the tensions.

3.5. Risk communication – a missing link?

There is one important aspect that, except for a couple of chapters, has been relatively underplayed in this volume. That is the role of the public and the consumer and their relation to the regulator. Reiman and Norros in their chapter do identify as one of the roles for the regulator communication with the public in such a way as to assure them that safety is being maintained. This role is not emphasised in other discussions of inspectorates but it is critical, they argue, in the case of nuclear power station inspectorates. The regulator effectively works for the public. Yet in many areas, the public is clearly uninvolved, and finds it difficult to follow or understand the regulatory process. There appears to be an increasing desire for such involvement however, as evidenced by increasing calls for public enquiries, and the coming into existence of many common interest groups, etc., who wish to influence safety directly or indirectly. This trend has undoubtedly been fuelled by the media, but nevertheless shows the perhaps natural interest of people in risks that can affect them. Public pressure, when it is mobilised, can be extremely powerful, as many companies and indeed whole industries have found out to their cost. However, understanding risks is not easy for the public, as risk communication seems to be largely either left up to companies who may often be optimistic about their risks, or the media, who are usually more pessimistic, both for reasons that are understandable, though not desirable. What is needed therefore, is better public risk communication. This should ideally be across different industries, allowing the public to make better (more informed) risk judgements. The regulatory authorities could play a significant role (as an unbiased 'broker' of information) in engendering better risk communication to, and ultimately with, the public.

A related issue which does come up in several chapter is the value of "naming and shaming" those who violate. This is seen as a powerful tool, but a two-edged one, which can destroy any sense of trust between regulator and regulated. It is often noted that the public demand for retribution following major accidents undermines the willingness of companies to identify their failings and learn from them, for fear that this will expose them to liability. Blame, in this sense, generates "organisational learning disabilities" (Baram, 1997).

4. Concluding Comments

This book has reviewed regulation in a number of industrial and public sector sectors. It has shown the diversity of regulatory styles and issues that exist and, above all, the complexity of the regulatory process. What is clearly needed with some urgency, is better understanding of regulation, and in particular of how to adapt regulation to evolving industries within fluctuating markets. This book and the workshop on which it was based can only scratch the surface of an issue as complex as safety regulation. This chapter therefore poses far more questions than it gives answers. Of the four questions which formed the basis for the workshop, we have only answered two in any detail. We have described a range of forms of regulatory control in use today and analysed how they have arisen and what those operating them believe to be their strengths and weaknesses. We have also indicated some of the ways in which they evolve. Our answers to the other two questions, about the matching of regulatory regimes to hazards and industries and about the degree to which we can learn from one regime or country in regulating another, are much more preliminary. We have offered some frameworks and tried to structure the influences and choices we have discussed. We hope that these can be used as a starting point for new research. The many questions the book poses indicate just how great the need is for evaluative work, which goes beyond the descriptive level at which much of this book is pitched. We need to know not just how regulatory regimes set about their work, but how successful they are in achieving their goals and what alternatives there may be which could be more effective.

There are also clear gaps in the coverage of the book. We have hardly touched on the issues of liability and damage claims, or of insurance. Both of these are complex issues interrelating with the regulatory approaches covered here. Sometimes and in some jurisdictions (notably the American) they have a dominant role in determining how risk is managed by society. In all jurisdictions organisations and processes concerned with liability and insurance are major players in carrying out a number of the steps in our problem-solving framework.

Regulation is currently seen as a valid societal function, but its added value is sometimes questioned by industry. The technological nature of many areas being regulated suggests that it needs to be more than a function. It needs to be seen as an approach, a discipline, and maybe even a craft[16], as suggested earlier. It needs to be seen as useful to industry and public sectors, not by being liberal and lenient, but by helping these sectors and industries to keep focused on the real threats they face.

There is a danger that government regulators could be replaced too unquestioningly by certification approaches. Certification has undoubted value and strength, but it can never take over completely from government. There are also dangers in too much delegation of regulatory tasks to the regulated, namely in self-regulation. If it goes too far regulators do not have the resources to check that all is well. In much of the discussion during the workshop and in preparing this book the issue of cost-effectiveness lurks just beneath the surface. There is a risk that pressures to slim down the government apparatus in order to save money can result in loss of skills and influence which are not picked up by other regulatory bodies. There needs to be some central optimisation of regulatory effectiveness against regulatory cost. Despite the many difficulties and complexities of government involvement in safety regulation, it would be potentially a huge mistake to write off the government inspector.

Guardians against risk, in one form or another, have pervaded human social structures for millennia. They are there because we know we make mistakes, individually and corporately. In this technology-driven world, with all its dynamism, interactiveness and hazard potential, now would not seem to the authors to be a good time to erode this important safety mechanism. However, although this chapter has argued that safety regulation is needed, the reader, and the public, have to make their own judgement on this issue, possibly based on consideration of how safe the world would be without safety regulation processes.

It is clear that this book has only made a modest beginning in trying to make sense of the regulation of risk from the perspective of safety science. There is a long way to go and much interesting work to be done both by researchers and practitioners in this field. We hope that this book stimulates many studies and perhaps helps to give them a coherent direction.

REFERENCES

Alders J.G.M. (2001). *Onderzoek cafébrand nieuwjaarsnacht 2001* (Investigation into the Café Fire on New Year's Eve 2001). Ministry of Internal Affairs and Kingdom Relations. Den Haag. Staatsuitgeverij.

[16] The use of the term 'craft' is intentional – a craft is something seen as useful to society, something that has a niche given the markets existing for that society.

Baram, M. (1997). Shame, blame and liability. In A. Hale, B. Wilpert, & M. Freitag (Eds.), *After the Event: From Accident to Organisational Learning*

Bardach, E., & Kagan, R. (1982). *Going by the Book: the Problem of Regulatory Unreasonableness*. Philadelphia: Temple University Press.

Berger, Y. (1999). Why hasn't it changed on the shopfloor? In C Mayhew & C Peterson (eds) *Occupational Health and Safety in Australia* (pp 52-64). Sydney: Allen & Unwin

Boden, L. (1985). Government regulation of occupational safety: underground coal mine accidents 1973-75. , **75**(5), 497-501.

Braithwaite, J. (1985). *To Punish or Persuade: Enforcement of Coal Mine Safety*. Albany: State University of New York.

Carson, W. (1970). White collar crime and the enforcement of factory legislation. **10**, 383-98.

Cullen, Lord. (1990). *The Public Inquiry into the Piper Alpha Disaster*. London: HMSO.

Cullen, Lord. (2001).

Fisse, B., & Braithwaite, J. (1993). *Corporations, Crime and Accountability*. Cambridge: CUP.

Gallagher, C. (1999). New directions: innovative management and safe place. *Proceedings of First National OHS Management Systems Conference, Sydney.*

Grabosky, P., & Braithwaite, J. (1986). *Of Manner Gentle: Enforcement Strategies of Australian Business Regulatory Agencies*. Melbourne: Oxford.

Gray, W., & Scholz, J. (1991). Analyzing the Equity and Efficiency of OSHA Enforcement. , **13**(3), 185-214.

Gray, W., & Scholz, J. (1993). Does regulatory enforcement work? A panel analysis of OSHA enforcement. **27**(1), 177-213.

Hale, A.R. 1978. The role of HM Inspectors of Factories with particular reference to their training. PhD Thesis. University of Aston in Birmingham. UK.

Hale, A., Baram, M., & Hovden, J. (1998). Perspectives on safety management and change. In A. Hale & M. Baram (Eds.), *Safety Management: The Challenge of Change* Oxford: Pergamon.

Hale A.R. & Hovden J. 1998. Management and culture: the third age of safety. In A-M Feyer & A Williamson (eds.) *Occupational Injury: risk, prevention and intervention.* Taylor & Francis. London pp 129-166.

Hale A.R. & Swuste S. 1998. Safety rules: procedural freedom or action constraint? *Safety Science*. **29** (3) 163-178

Hopkins, A. (1994). Compliance with what? The fundamental regulatory question. *British Journal of Criminology*, **34**(4), 431-443.

Humphreys, P. [Ed.] (1988) Human Reliability Assessor's Guide. Report SRDA-R11, AEA Technology, Culcheth, Warrington, UK.

Hutter, B. (1989). Variations in regulatory enforcement styles. *Law and Policy*, **11**(2), 153-174

Kagan, R., & Scholz, J. (1984). On regulatory inspectorates and police. In K. Hawkins & T. J.M. (Eds.), *Enforcing Regulation* Boston: Kluwer-Nijhoff.

Lamm, F. (1999). Small business: An overview of a NOHSC project. *Paper to the ANZAOHSE Conference, February , Auckland.*

Larsson T.J. & Clayton A. (eds.) 1994. Insurance and prevention: some thoughts on social engineering in relation to externally caused injury and disease. IPSO. Stockholm.

Mayhew, C. (1999). Why owner/managers in small business miss out. In C. Mayhew & C. Peterson (Eds.), *Occupational Health and Safety in Australia.* Sydney: Allen and Unwin.

Neutze, M. (1997). *Funding Urban Services.* Sydney: Allen & Unwin.

Oosting 2001

Perry, C. (1982). Government regulation of coal mine safety: effects of spending under strong and weak law. **40**, 303-314.

Rasmussen, J. (2000). Safety – a communication problem? Paper to the NeTWork Workshop on Safety regulation. Bad Homburg

Rasmussen J. & Svedung I. (2000). Proactive risk management in a dynamic society. Swedish Rescue Services Agency. Karlstad, Sweden.

Reason, J. (1990) *Human Error.* Cambridge University Press, Cambridge.

Turner, B. (1978) *Man-made disasters.* Wykeham, London.

Wells, C. (1993), *Corporations and Criminal Responsibility.* Oxford: Clarendon

KEYWORDS